CAMBRIDGE COUNTY GEOGRAPHIES

SCOTLAND

General Editor: W. Murison, M.A.

AYRSHIRE

Cambridge County Geographies

AYRSHIRE

by

JOHN FOSTER, M.A.
Headmaster, Academy, Beith

With Maps, Diagrams and Illustrations

Cambridge:
at the University Press
1910

CAMBRIDGE UNIVERSITY PRESS
Cambridge, New York, Melbourne, Madrid, Cape Town,
Singapore, São Paulo, Delhi, Mexico City

Cambridge University Press
The Edinburgh Building, Cambridge CB2 8RU, UK

Published in the United States of America by Cambridge University Press, New York

www.cambridge.org
Information on this title: www.cambridge.org/9781107634688

First published 1910
First paperback edition 2013

A catalogue record for this publication is available from the British Library

ISBN 978-1-107-63468-8 Paperback

CONTENTS

ILLUSTRATIONS

MAPS

The illustration on p. 36 is reproduced from a photograph by Mr Charles Kirk, Glasgow; that on p. 161 from a photograph by Messrs Irvine & Son, Kilmarnock; and those on pp. 126, 127 from photographs by Mr Alexander D. Henderson, Maybole.

The author is also indebted to Mr James Stevenson, Beith, for the views on pp. 24, 121 and 122; to Mr John Stevenson, Beith, for that on p. 110; to Mr James Newbigging, West Kilbride, for that on p. 48; to Mr Adam M'Arvail, Mauchline, for that on p. 165; to Messrs William Fife and Son, Fairlie, for the use of the illustration on p. 44; to Nobel's Explosives Company for that on p. 78; to Messrs Morton of Darvel and Messrs Virtue & Co. for use of that on p. 76; to Glengarnock Iron Company for the use of those on pp. 79 and 80; to the Society of Antiquaries of Scotland for permission to reproduce those on pp. 98, 99, 100, 101, 102, 103, 104, 107; to Mrs Marian Murdoch for that on p. 146.

1. County and Shire. The name *Ayr*.

With respect to the subject of this volume, the names *shire* and *county* are interchangeable. Though the terms are practically synonymous, they are not arbitrary distinctions, for each has a characteristic signification which conveys lessons bearing on the social condition of our county's remote past. *Shire* in its etymology is allied to *share* and *shear* and is properly a portion *shorn off*. For military and financial purposes the old kingdoms were broken up into divisions, each of which was required to furnish a fixed number of men in the event of war and to levy an assigned amount of money as its contribution to the national necessities. The government of this *share* or *shire* was assigned by the Saxon kings to an earl or alderman, and, in order to enforce its obligations to the crown, the jurisdiction of the district was entrusted to an official deputy called the "shire-reeve," the modern sheriff. After the Norman Conquest the Saxon *earl* was displaced by a nobleman of similar rank who came across with the Conqueror, and from the fact that he had had the honour of being chosen close companion to his leader he was called *Comes* (Latin *comes* = a companion). He had duties largely corresponding to those of a lord-lieutenant of the present time. It was his business to rule one of the local divisions formerly existing—*comitatus*—and from this Latin designation the English word *county* ultimately came.

For many of these divisions in the British Isles the double nomenclature is in use; with others only one term is usually associated. While all of them may be appropriately called *county*, the name *shire* can be attached to only some of them, the reason being that these are parts of larger divisions called kingdoms. Kent, for example, never goes by the name of *shire* because it represents the old kingdom of *Cantium*. Cornwall, too, is the land of the *Welsh at the Horn*, and for that reason it is only known under the designation of *county*. Ayr, inasmuch as for several centuries it formed part of the kingdom of Strathclyde, acquired the name of *shire*.

How was the county named *Ayr*? As in many other divisions, it received its name from the chief town, which in turn was named from the river on which it stood. *Ar* from which the word is derived is a root widely diffused both in the British Isles and in continental countries, and a twofold meaning is associated with it. It may either signify *slowness*, as in the *Arar*, which, according to Caesar, flowed with a smoothness beyond belief, or it may imply the opposite quality, *swiftness of motion*, as in the Is*ar* "flowing rapidly." With respect to the Ayr the balance of opinion seems to incline to the former of the two meanings because the bed of the river, in the main, lies on a smooth even stratum, giving the water a leisurely movement. The other feature is, however, equally characteristic of portions of the stream, where, to the partial view of our national bard, it was correctly described as "bickering to the sea."

2. General Characteristics. Position and Natural Conditions.

Ayr is a maritime county, the broad estuary of the Clyde washing its western shores. Far out lies the busy ocean highway for vessels from and to the bustling commercial centres in the upper reaches. Its coasts are indented with excellent harbours. Its shores, to a large extent, are sandy and many of its towns are now popular sea-side resorts. Over much of their broad waste covered with bent grass, hard and wiry, stretches a long series of golf-links, 21 in number, some of them the most famous in the world. Among the rocks in the south are several remarkable caves. On its eastern boundary the whole district may be said to be shut in from the adjacent counties by high ridgy land, and the south is a hilly and wild region throughout.

In some respects Ayrshire is one of the most representative counties in the British Isles. All its interests are very evenly balanced, whether these are maritime, agricultural, industrial, or mineral.

Before the abolition of feudalism the county was divided into the three districts (still popularly recognised though without official meaning) of Cunningham in the north, Kyle in the centre, and Carrick in the south. Before the development of the industries and minerals of the district the characteristics of each of these localities were supposed to be hit off in the quaint local rhyme of unknown antiquity:—

"Kyle for a man: Carrick for a coo:
Cunningham for butter and cheese;
And Galloway for 'oo."

The ancient bailiwick of Cunningham, extending from the Irvine water to the northern boundary of the county and comprising 16 parishes, was governed hereditarily by the great family of Cunningham, the head of which bore the title of Earl of Glencairn. Kyle, lying between the Doon and the Irvine and with 21 parishes, was divided by the river Ayr into Kyle Regis or King's Kyle in the south and Stewart Kyle in the north. Stewart Kyle, or part of it, was the property of the old hereditary Stewards of Scotland, who by marriage became the Royal Stewarts. Carrick, the division that extends from the Doon to the southern boundary, comprises 9 parishes. It is in general mountainous with some pleasant valleys interspersed. The district gave the title of Earl to Robert Bruce of Turnberry Castle in this district, and the present bearer of the title is his descendant, the Prince of Wales.

A considerable extent of Ayrshire from the foot of the Doon to near West Kilbride in the north is a plain open tract, only a small part of which can be termed level. There are numerous swells which facilitate the escape of moisture and promote ventilation, and a gradual acclivity terminates in rugged mountains in the south-east and in moorish hills in the east and north.

The great vale of Cunningham and the coast lands of Kyle are in the main fertile while the north-western angle of Carrick and a strip in the west are the only parts of that division that are of much agricultural value. While Ayrshire occupies a foremost place in agricultural husbandry, the working of its dairy lands has reached a high degree of excellence, and is superior to that of any other district of Scotland. The ready means of communication by which the county is everywhere so well served renders the interchange of products expeditious and cheap.

The trail of industry is much in evidence throughout the two northern divisions, where the tranquil meadow, the gliding river, the greenwood, and the cornfield are nowhere far removed from the whir of wheels and the din of traffic. Manufactures have attained considerable importance and the district possesses great advantages for their development. The mineral riches of the county are considerable, and building materials exist in large quantities and are well distributed.

Although no epoch-making battle has perhaps been fought within its border, if we except that against King Haco in 1263, which put an end to Scandinavian aggression, the county can number among its stirring incidents many of national interest. Not a few of the exploits of Wallace and Bruce in furthering the cause of national independence were achieved in this district. Besides, it may be doubted if any other county has had more exciting events than those of its family blood-feuds, which tore it so disastrously and so long defied the authorities to quell. The protracted struggles, too, on behalf of civil and religious reform, of which this county was the centre, were fraught with significant and eventful issues.

The scenery is varied and picturesque. The south is wild, craggy, and gloomy with deep dells, narrow ravines, and rocky gorges, and the scenery is greatly influenced by the character of the rock forming the rugged mountain masses. As the wayfarer tops one of the eminences in the north and gets his first glimpse of the sea the picture before him can hardly be surpassed. The Firth is at his feet, its surface mottled with islands. The profile of jagged peaks rising dark and high fills the background. The channel is dotted with pleasure craft and rich merchantmen, and the lone sea-girt dome of Ailsa looms in the offing. From another eminence of no great elevation, also in a northern parish, a magnificent panoramic view is obtained, commanding as it does a large part of the Firth of Clyde, while most of the lowlands of Ayrshire are spread below like a map unrolled, with the mountain ranges of Galloway and Carrick, the Heads of Ayr, the craig of Ailsa, and the lofty peaks of Arran bounding the horizon to the south and west, while northward is observed the rugged ridge of Cowal and the wavy outline of the far-off hills of Perthshire with Ben Lomond standing prominently in front.

3. Size. Shape. Boundaries.

The greatest length of the county measured in a direct line from north to south is 55 miles. The line of coast, however, extends to 84 miles. The breadth varies from five miles at either extremity to 26 miles at the widest part of the crescent measured through a point immediately south of Troon. If the zigzags of the inland boundary line are all included from Galloway Burn in the south north-east along the mountain-tops by Loch Doon and Corsancone Hill to Muirkirk, and from that point north-west along the hill-tops by Loch Goin to the Kilbirnie Hills as far as Kelly Burn, the length is no less than 160 miles. The area amounts to 735,262 acres, or almost 1150 square miles. Among Scottish counties Ayrshire ranks seventh in size, being about one-fourth of Inverness, the largest, and thirty times greater than Clackmannan, the smallest. Compared with the adjacent counties, it is four and a half times the size of Renfew on the north; one-third larger than Lanark, slightly greater than Dumfries, one-fifth more than Kirkcudbright, in the east; and two and one-third times the size of Wigtown in the south.

In shape Ayrshire is an irregular crescent, the trend of the coast curve being mainly eastward over half its length. It then turns to the south-west and keeps that course until Loch Ryan at the southern extremity is reached. Three small promontories disturb this general tendency of the coast-line. One of these, the foreland of

Portincross, including the celebrated precipice of Ardneil Bank and Farland Point, abuts into the Firth about ten miles south of the northern boundary. About five and a half miles south-east is the rocky spit on which Ardrossan stands, while nine miles further in the same direction is the rugged horn on which a large part of the town of Troon is built.

Only on its western or concave side has the county a natural boundary—the Firth of Clyde and the North Channel. On the convex side the frontier is defined by more or less arbitrary and artificial landmarks depending less upon physical features, though, as far as possible, streams and lochs are made to serve as the boundary-line. It is enclosed by high ridgy land on the north and east, while on the south-east a stretch of rugged tumbling mountains separates it from the adjoining counties already named. For short distances stretches of water mark the border. Separating portions of the county from Renfrewshire are Kelly Burn, the Maich, the Dubbs, Roebank Burn, and Loch Goin. For part of its course the Avon flows between Ayrshire and Lanarkshire, and the Kello Water separates the same county from Dumfriesshire. Kirkcudbrightshire shares with it the Deugh Water, Loch Doon, Eglin Lane, and the river Cree, while Loch Dornal, Loch Maberry, Pulganny Water, Cross Water of Luce, Pulhatchie Water, Drumorawhirn Water and Galloway Burn, form part of the frontier between it and Wigtown.

4. Surface and General Features.

A reference to the map will show that the surface of Ayrshire admits broadly of three divisions—the high ridgy land that forms the landward boundary of the county in the east and north, the broad undulating fertile plain stretching from the northern march 40 miles south to Girvan with a broad strip of fertile ground in the upper valley of the Nith, and the wild hilly region of Carrick.

The upland district in the north stretching from the shore of the Clyde estuary north-east into Renfrewshire is known as the Kilbirnie Hills. A prominent feature in the foreground on the county boundary is Misty Law (1663 feet), so named "because it is almost ever covered with dark mists and thick fogs." Also on the boundary, not so well seen because further back from the edge of the ridge, rises Hill of Staik, 50 feet higher. Several other minor eminences here and there break the monotony of this stretch of upland. These hills are all grass-topped, and, with the exception of some clumps of plantations along the slopes, for the most part they are scantily wooded. The green hill-sides, however, towards the Firth are well timbered. Misty Law, there is reason to believe, was at one time covered with coniferous trees.

The belt of high land that fences off the eastern frontier of Ayrshire from the adjacent counties is of a varied character. In the north the hills are neither wild nor picturesque, but, stretching the full extent of the county, with three depressions to interrupt their course, they become grander towards the south and culminate in a region of rugged magnificence. The only eminences of any note in the north before Loudoun Hill is reached are Black Law and Crookhill. Loudoun Hill is an attractive feature of the landscape not so much on account of its elevation as from its isolation. It attains the height of only 1034 feet above the sea-level, or 600 feet above the

Loudoun Hill

level of the surrounding country. Standing in one of the depressions already referred to it is seen from a great distance all around. It has been described as "rising up in the air all alone by itself and that out of a very low soil." The hill is of a conical shape and basaltic formation, and is supposed to form part of a large trap dyke which cuts the whole coalfield of Ayrshire in a north-west and south-east direction. From Loudoun Hill southward the hills take a more mountainous character. Distincthorn (1258 feet) is the first of the more pretentious eminences. Cairn Table (1942 feet), a well proportioned mountain, Corsancone Hill (1547 feet), consecrated by Burns in one of his lyrics as the Scottish Parnassus, the Knipe Hills (1950 feet), Blackcraig (2298 feet), so called from the scarcity of vegetation on its blue-grey western front, Blacklarg (2231 feet), and Windy Standard (2287 feet) bring us to a desolate region of crags and gullies where Shalloch-on-Minnoch rises to the height of 2520 feet. This last is backed by Kirrierioch Hill (2562 feet) and the spur terminates in one of the most elevated hills of the district, Mount Merrick (2764 feet), situated just over the county frontier.

Many other peaks of considerable elevation are scattered over several of the southern parishes. Knockdolian is an isolated eminence near the coast, its shoulder terminating in Bennane Head. Further north, and also near the coast, is Brown Carrick Hill (940 feet), a pastoral height, its broad base abutting into the sea and forming the rugged foreland of the Heads of Ayr, a conspicuous mark on the coast. In the south of Ochiltree is Stannery Knowe (1191 feet), and Craigengower (1086 feet) overlooking Blairquhan Castle is crowned with a monument to Colonel Blair of Blairquhan. North of Dalmellington is Benbeoch (1521 feet), affording a grand view of New Cumnock vale and hills. Enoch Hill (1865 feet) is the source of the Nith. About eight miles west of Corsancone Hill is the lofty moorland wild of Corsgalloch (1192 feet), on which stands a monument to martyrs during the covenanting struggle. About the same distance further west the Craigs of Kyle rise by a gentle acclivity to the height of 799 feet, and from the top splendid views are obtained all around.

Along the shore in the south of Cunningham there is an extensive tract occupied by dunes or sand-hills, which

Benbeoch Craig, Dalmellington

the winds blowing over the broad sea-beach have piled up there. These are called the Ardeer Sands. Almost the whole area of this waste is now occupied with the world-renowned Nobel's Explosives Factory, which for the sake of safety isolates its various departments among the scattered dunes.

Here and there throughout the county large tracts of

moss occur, darkening the landscape. The largest of these are Calder Moss, lying to the north-west of Loudoun Hill, and Aird's Moss, a wild dark region in the district of Kyle between the Water of Ayr in the north and Lugar Water in the south. This uncultivated stretch of bleak mossy ground is about five miles long by two broad. At its head stands a monument to the memory of Richard Cameron, famous in the annals of the Scottish covenanting struggles. Riccarton Moss, near Kilmarnock, is of several hundred acres in extent. Large flow mosses are also met with in the upper end of the parish of Kilwinning and the lower parts of Stewarton and Beith, and that of Shewalton or Auchans near Irvine extends to within a short distance of the shore. All this bogland, consisting on the surface of bare heather stems warped and bleached by wind and rain, is gradually being reclaimed and cleared into profitable lands with, at first, a rank growth of poor grasses, and afterwards a rich cultivation of suitable crops.

The less rugged hills on the county boundary in the south-east clad with bent and heath are the best examples of the Ayrshire sporting moors where grouse, partridge, and black-game are the main objects of pursuit. Mossy moors and hilly wastes in the east of Kyle, as well as benty hills in the south and north above the limits of cultivation, are also well adapted to this form of sport.

If man, as is said, made the town, he is also responsible to a considerable extent for defacing large tracts of the country. Our county, it is to be regretted, has in places some ugly blots upon its fair landscape. Heaps of slag, blaes "bings," and large black masses of coal dust that have accumulated round the blast furnaces and collieries raise their bald heads and render their neighbourhood unsightly. Happily these are in process of disappearing, the waste of one industry being profitably used as the raw material of another. Should this operation continue the archaeologist of the remote future, investigating the tumuli of the past, will be relieved of any puzzling enquiry as to what these mounds might have meant.

5. Watersheds and Rivers.

The high ridgy and mountainous land forming the convex boundary of the county makes a very well defined watershed where all its rivers rise and whence all but one bear westerly their fruitful streams. The Stinchar, the Girvan, the Doon, and the Ayr with the Lugar mainly drain the southern half of the county, while the Irvine and the Garnock are the chief arteries of the northern division. In addition to these many a thread-like stream strays round rocky scaur, wimples through a glen, or twists with many a turn in the lap of the hills.

None of the rivers are navigable. They are, therefore, of little commercial value except for the salmon and trout with which some of them abound. The estuaries of the Ayr and the Irvine, however, lend themselves to the formation of excellent harbours.

For a short distance along the southern boundary the Galloway Burn flows west and empties itself into Loch

Ryan, into which, a mile northward along the pleasant shore, falls also App Water.

The Stinchar rises in a small lake high in the moorland parts of Carrick and flows at first with a very rapid course. In its middle and lower reaches through Colmonell and Ballantrae it glides with slower current, drains some well-wooded, picturesque plains, and joins the Atlantic at

View on the Stinchar, Ballantrae

Ballantrae, 30 miles from its source. The Stinchar is reinforced by several streams, particularly by the Water of Assel on its right bank, and by the Dusk and the Tig on its left.

Girvan Water rises 1500 feet above sea-level in a small loch called Girvan Eye in the south of Straiton parish, and at only a short distance from the head waters of the Stinchar. At first its course lies through a succession of lochs—Cornish, Skelloch, Lure, and Bradon. Owing to the declivitous character of its bed, at first, it descends rapidly, earning its name (Girvan means rapid river) as it tumbles over the rocks. It proceeds in a north-west direction through wild moors until it reaches Tarelaw Linn, where it dashes over cliff after cliff and rages through a narrow gorge half a mile in length. Thence passing through an open vale, and turning south-west, it rolls forward through a richly wooded valley past Dailly; then with slower current it reaches the sea six miles further on at the town of Girvan. Above Dailly it receives on the left the waters of Shiel Burn, which traverses a valley finely diversified with wooded slopes.

Loch Doon, a dark sheet of water six miles long and at its widest part three-quarters of a mile broad, lies between high moors among the rugged hills in the west of Carrick. The northern half lies wholly in the county, the southern half forms part of the boundary between Ayrshire and the county of Kirkcudbright. Numerous other small lakes lie sparkling near and drain into Loch Doon. Its main supply, however, is from Loch Enoch, seven miles south on the Kirkcudbrightshire border, from which issue into Loch Doon Gala Lane and Eglin Lane. Loch Doon is the largest lake in the county and has attained some celebrity for its trout fishing.

From the north end of the loch issues the river Doon, the beauty of whose "banks and braes" has been immortalized by Burns. From source to mouth this river

Head of Loch Doon

forms the boundary between Kyle and Carrick. Leaving the loch it descends at once into the bosom of Ness Glen, an amazingly narrow and deep ravine, one of the grandest natural objects to be seen in Ayrshire. Pent up in a channel only four or five yards wide, the river churns among the rocks, cliffs rising perpendicularly on either side to the height of 230 feet. Emerging from the glen and Berbeth woods it expands into a reedy loch, then veers from a north to a north-west direction for some miles, flowing sluggishly through bogs and meadows past Waterside with its ironworks on the right bank three miles below Dalmellington. A little further down the river, Patna and Carnochan are passed on the left. The valley is now bounded by the fairy-haunted Cassillis Downans, celebrated in *Hallowe'en*. Its slopes again become graced with woods as the river proceeds past Dalrymple and Alloway. Less than a mile further it reaches the sea at the southern extremity of Ayr Bay after a course of 22 miles.

A glance at the map shows that there is a small tract of the county lying to the north-east of Dalmellington with a drainage system peculiar to itself, its western watershed consisting of a series of moorish hills running nine miles north and north-east from that town and parting the infant streams of the river here on the east from those of the Doon on the west. This district is drained by the Nith and its tributaries. The Nith has its source about six miles due east from Dalmellington near the confines of Kirkcudbright, and flows first north, then east past the town of New Cumnock, which lies

Dalcairnie Linn, Dalmellington

scattered about the confluence of this river with its tributaries of Afton Water from the south and Moorfoot Burn from the north. It then cuts its way eastwards through a valley where the line of hills is cleft, and, after a number of windings of singular beauty called the "Loops o' Nith," it reaches the county march at the foot of Corsancone. This hill gives the course of the river a south-easterly direction and sends its waters through Dumfriesshire to the Solway. Its total length is 53 miles of which only 12 are in Ayrshire. Afton Water, the chief tributary of the Nith, rises among mountains on the very frontier of Kirkcudbrightshire, and in its upper course flows through a moorland valley past Montraw Burn to Castle William Falls. It then descends for a mile and a half among boulders between rocky banks to Craigdarroch at the foot of the frowning Blackcraig Hill. All this romantic district Blind Harry, the Minstrel, associates with the movements of Sir William Wallace during the seven years he was in comparative obscurity. The classic stream continues its course, murmuring "among its green braes," and "flowing gently" as the hills become less steep.

The Ayr, rising at Glenbuck in the eastern extremity of the county, flows mainly west, dividing it into two nearly equal portions. After receiving many additions from the drainage of the central plain and passing on its way the towns of Muirkirk, Sorn, Catrine, Stair, and Annbank, it loses itself in the Firth of Clyde at the town of Ayr, where its estuary forms the harbour. It is for miles of its course only a small rivulet flowing through an open moorland district, but being joined on the right by the Greenock Water and the Cleugh Burn, and on the left by the "haunted" Garpel, the "winding" Lugar, and the "brawling" Coyle, it becomes a large body of water. Over much of its lower course it is bounded by steep, rocky, wooded banks, between which it grandly winds its way downwards. This is especially the case at Ballochmyle and Barskimming, where we have typical Ayrshire scenery. Here and there the banks open and some delightful haughs occur, but in many places the river is seen fretting through long, deep, narrow chasms overhung by luxuriant foliage. After continued rains in the upland districts the river is subject to heavy floods which inflict serious damage on the low-lying grounds. It must have been this circumstance that suggested to Burns the descriptive touch in depicting the familiar stream as "one lengthened tumbling sea."

The Irvine rises in two head waters, the one in a moss at Meadowhead on the eastern boundary of the county, and the other a mile further east in Lanarkshire near the battlefield of Drumclog. If its numerous windings are neglected the river pursues a uniform westerly course until near the place where it enters the sea. About two and a quarter miles westward from the Ayrshire boundary it is joined from the south by Logan Water, and half a mile onward by Glen Water from the north, which on account of its size ought strictly to be the parent stream. Near the town of Irvine the river makes a bend like one of the Links of Forth, assuming a southerly direction immediately after a fine sweep northward. It abruptly expands into a basin three-quarters of a mile broad, and communicates by a narrow strait with the Firth of Clyde. Many features combine to make it one of the most pleasing of rivers—the richness of its haughs, the openness of its course, the quality of the adjacent soil, the display of industry everywhere observable along its course, and the numerous residences of note that overlook the river and its tributaries.

The Garnock (Celt. *garbh* = rough, *cnoc* = a knoll) rises at the base of the Hill of Staik. About a mile and a half from its source it tumbles noisily over a rocky bed of porphyry 60 feet high, and forms a wildly picturesque cascade called the Spout of Garnock. Then it winds round two sides of the precipitous knoll on which are perched the tottering ruins of Glengarnock Castle. With the exception of the immediate environs of the Spout of Garnock and Glengarnock Castle the scenery for the most part is tame and uninteresting. It enters the sea at Irvine. One of the tributaries of the Garnock is the Rye, on which the most interesting spot is the ford above Dalry. Some amusing scenes at the ford are depicted in the familiar song *Gin a body meet a body*. Another tributary is the Dusk Water. A short distance above the confluence is a stalactite cave called Cleaves Cove, one of the greatest natural curiosities in Ayrshire. The New Statistical Account gives the following description of it—"On the estate of Blair in the romantic and beautifully wooded glen of the Dusk there is a natural cave in a precipitous bank of limestone. It is about 40 feet above the bed of the stream, and is covered by about 30 feet of rock and earth. It has two entrances. The western, or main entrance, is situated below a vast overhanging rock, 30 feet long by 27 in breadth....Its interior resembles Gothic arched work. Part of the roof is supported by two massy columns. Its length is about 183 feet, by 12 broad, and 12 high. Its internal surface is covered by calcareous incrustations, and numerous crevices branch off from its sides. In former times popular belief peopled it with elves. It consequently acquired the name of the 'Elf House.' In later days, during the tyrannical reign of Charles II, it afforded a hiding place to the covenanters from the violence of their infuriated persecutors."

Flowing parallel with the upper reaches of the Garnock, and little more than a mile east of it, is the Maich Burn, which, rising in Misty Law Moor in Renfrewshire, falls into the northern extremity of Kilbirnie Loch, forming part of the boundary between Ayrshire and Renfrewshire. Kilbirnie Loch, "the goodliest freshwater loch in all Cunningham," is one and a half miles long and less than half a mile broad. In general its banks are low and shingly and almost devoid of timber. It, therefore, little lends itself to picturesqueness. Its surface is 95 feet above the level of the sea. The limit of the watershed of the long valley stretching from near Paisley to Irvine harbour lies between the southern extremity of the loch and the Garnock close by, although there is no apparent intervening elevation to form it. The waters of the loch flow north by the Dubbs into Castle Semple Loch, and thence by the Black Cart to the White Cart at Renfrew, where it joins the Clyde. A mile down from the loch the Dubbs is joined on the right by Roebank or Rough-bank Burn, which also rising in Renfrewshire forms for most of its course a natural boundary between the two counties.

Loch Doon and Kilbirnie Loch already noticed are

Roebank Burn, near North Lodge, Woodside, Beith

the largest lochs in Ayrshire. There are others, some of considerable extent, scattered over the county, especially in the mountainous parts in the south. There are no fewer than 17 within three or four miles of Loch Doon, and 20 more in other parts of Carrick, 16 in Kyle, and four or five in Cunningham. The chief are Lochs Finlas, Riecawr, and Macaterick "set dark and deep in treeless wastes" near Loch Doon, into which they drain. Lying at no great distance from the Craigs of Kyle are three lochs—Loch Kerse, Loch Fergus, and Loch Martnaham. Of these the last is by far the largest, and in the middle of it is an island with the ruins of one of the ancient Ayrshire keeps.

6. Geology and Soil.

The rocks are the earliest history books that we have. To those who understand them they tell a fascinating story of the climate, the natural surroundings and the life of a time many millions of years before the foot of man ever trod this globe. They tell of a long succession of strange forms of life, appearing, dominating the world, then vanishing for ever. Yet not without result, for each successive race was higher in the scale of life than those that went before, till man appeared and struggled into the mastery of the world.

The most important group of rocks is that known as *sedimentary*, for they were laid down as sediments under water. On the shores of the sea at the present time we find accumulations of gravel, sand, and mud. In the course of time, by pressure and other causes, these deposits will be consolidated into hard rocks, known as conglomerates, sandstones, and shales. Far out from shore there is going on a continual rain of the tiny calcareous skeletons of minute sea-animals, which accumulate in a thick ooze on the sea-floor. In time this ooze will harden into a limestone. Thus by watching the processes at work in the world to-day we conclude that the hard rocks that now form the solid land were once soft, unconsolidated deposits on the sea-floor. The sedimentary rocks can generally be recognised easily by their bedded appearance. They are arranged in layers or bands, sometimes in their original horizontal position, but more often tilted to a greater or less extent by subsequent movement in the crust of the earth.

We cannot tell definitely how long it is since any special series of rocks was deposited. But we can say with certainty that one series is older or younger than another. If any group of rocks lies on top of one another then it must have been deposited later, that is it is younger. Occasionally indeed the rocks have been tilted on end or bent to such an extent that this test fails, and then we must have recourse to another and even more important way of finding the relative age of a formation. The remains of animals and plants, known as fossils, are found entombed among the rocks, giving us, as it were, samples of the living organisms that flourished when the rocks were being deposited. Now it has been found that throughout the world the succession of life has been roughly the same, and students of fossils (palaeontologists) can tell by the nature of the fossils obtained what is the relative age of the rocks containing them. This is of very great practical importance, for a single fossil in an unknown country may determine, for example, that coal is likely to be found, or perhaps, that it is utterly useless to dig for coal.

There is another important class of rocks known as *igneous* rocks. At the present time we hear reports at intervals of volcanoes becoming active and pouring forth floods of lava. When the lava has solidified it becomes an igneous rock, and many of the igneous rocks of this country have undoubtedly been poured out from volcanoes that were active many years ago. In addition there are igneous rocks—like granite—that never flowed over the surface of the earth as molten streams, but solidified deep down in subterranean recesses, and only became visible when in the lapse of time the rocks above them were worn away. Igneous rocks can generally be recognised by the absence of stratification or bedding.

Sometimes the original nature of the rocks may be altered entirely by subsequent forces acting upon them. Great heat may develop new minerals and change the appearance of the rocks, or mud-stones may be compressed into hard slates, or the rocks may be folded and twisted in the most marvellous manner, and thrust sometimes for miles over another series. Rocks that have been profoundly altered in this way are called *metamorphic* rocks, and such rocks bulk largely in the Scottish Highlands.

The whole succession of the sedimentary rocks is divided into various classes and sub-classes. Resting on the very oldest rocks there is a great group called Primary or Palaeozoic. Next comes the group called Secondary or Mesozoic, then the Tertiary or Cainozoic, and finally a comparatively insignificant group of recent or Post-Tertiary deposits. The Palaeozoic rocks are divided again into systems, and since the rocks of Ayrshire fall entirely under this head, we give below the names of the different systems, the youngest on top.

Palaeozoic Rocks.

Permian System.
Carboniferous System.
Old Red Sandstone System.
Silurian System.
Ordovician System.
Cambrian System.

In a general way the rocks of Ayrshire may be ranged in three divisions—Carboniferous in the north, Old Red Sandstone in the middle, and Lower Silurian with trap in the south. While this may be stated roughly, the geological systems are often very complex, for a great development of Old Red Sandstone and Carboniferous strata occupies large stretches of the county as far south as Dalmellington. Whinstone, Greenstone, and Red Sandstone are found in Carrick, while unmixed granite of a greyish colour abounds in the upper parts of the same district. What first strikes one with respect to the statement just made is that the rocks of the county are arranged according to their age in regular order from south to north, the older formation being in the south. This variety of rock gives wide differences in natural features from the picturesque scenery of the older rocks contorted and raised far above their original position to the comparatively monotonous levels of the alluvial plain. The hills in the south and south-east, mainly of granite formation pierced by igneous intrusions, being better capable of resisting destructive agencies, rise in a rugged grandeur which imparts a pleasing variety to the county.

Next in point of age as well as in point of order in the county is the Old Red Sandstone band. On the south-east of the great central depression in Scotland the Old Red Sandstone, after skirting the Lammermoor Hills, strikes south-west and reaches the sea south of Ayr. Traces of it are to be seen from Blairquhan in the Girvan valley to the shore at the Heads of Ayr. It is also to be observed near the sea in the northern parts of the county from Skelmorlie southward by Largs and Fairlie to Ardrossan. Ardneil Bank consists of dark Red Sandstone lying horizontally, and the same strata make their appearance at different places near West Kilbride.

Next come the Carboniferous rocks, a name applied to a great series including calciferous or yellow sandstone, carboniferous limestone, upper or mountain limestone, millstone grit, and the coal measures. The yellow sandstone forms a kind of passage from the Old Red Sandstone to the Carboniferous rocks proper, and its strata rests either on Silurian beds as at Kilkerran or on Old Red Sandstone as at Sorn. In the northern hills, stretching south to Ardrossan and south-east to Loudoun Hill plateau, eruptions are traceable in the porphyritic lavas met with along some of the streams, and on Ardneil Bank beautiful brown porphyry surmounts the red sandstone. On the volcanic débris rests the lower carboniferous series which is best developed in the Beith and Dalry districts. Like the lower limestone series the upper or mountain limestone—the light blue stone with which we are all familiar

—also attains its full development in the north, the very thickest lying to the south-east of Beith. It generally contains more clay and is seldom burned. On the other hand it is an excellent material for road-making. Another great division of carboniferous rock is the millstone grit, so named because it is the stone from which millstones are made, being a coarse quartzose sandstone much esteemed for hardness and durability. These are quarried to some extent near West Kilbride. The coal measures are another important division of the same rocks, and run through the centre of the county from the shore to its inland verges. Between the lower and the upper limestone series occur the lower coal and ironstone series, bands of the latter generally occupying the greater depths. Around Dalry there run extensive seams of the *blackband* ironstone, which among Scottish miners is supposed to contain coaly matter sufficient for calcining the ore without the addition of coal. These seams, however, have sometimes been found difficult to reach owing to the existence of intrusive igneous rocks called dykes which have to be worked through. The upper coal and ironstone series occupy much of the county, and are associated with the usual clays, shales, and white and grey sandstone of the division. The workable seams range from three or four in the Dalry district to seven or eight at Dalquharran, and near New Cumnock one of these was 40 feet thick. Succeeding the carboniferous formation and with greater affinity to it than to the New Red Sandstone, is the Permian. This is well exposed on Lugar Water near Auchinleck, and in several large quarries near Mauchline.

In different parts of the county are traces of extinct Permian volcanoes. Rocks of volcanic agglomerates mark the sites of these south-east of Symington, east of Irvine, and near Stevenston.

Remains and traces of living organisms preserved in rocks are called fossils. These may also consist of the cast or impression of some object left on an ancient surface of clay or wet sand, and the term is even extended to worm-tracks and ripple marks often observed on the soft sands of a shelving beach, to footprints, and to rain-pittings on soft rocks. Among Silurian and carboniferous fossils, found in great abundance in Logan Water just over the county border in Lanarkshire, are great water scorpions from the shale beds. Plant remains, ferns, seed-bearing plants, and giant reeds have frequently been discovered in the coal deposits. Carboniferous limestones being sedimentary deposits brought by oceanic currents are richly charged with marine organisms. On Mulloch's Hill north of Dailly a great variety of trilobites, corals, and shells have been found. In the bed of Old Red Sandstone near West Kilbride occur footprints probably reptilian, worm trails, and rain-pittings, and on the rocks on the shore between the mouth of the Doon and Heads of Ayr fish and plant remains are occasionally met with.

Many evidences go to show that in the remote past our islands were in great part covered with glacier ice and that there were periodic recurrences of an Arctic climate. Although it can be shown that large areas have been moulded by ice, the remains of ice action are singularly scanty over the county. Occasionally at the shore level, as at Ardneil Bay near West Kilbride, a glacial formation occurs composed of tough, reddish boulder-clay or drift. The same formation has also been observed in the Garnock Water and the Water of Irvine. Glaciers bring down with them into the valleys the débris of rocks, and, when these rocky fragments reach the end of their journey, owing to the melting of the ice they are left as huge mounds, called terminal moraines. Examples of these formations are to be found—one at the Cuff Hill, north of Beith, and one in the neighbourhood of Colmonell. In the bed of the Roebank Burn, on the Renfrewshire border, lies an erratic boulder supposed to have been transported by ice agency from the Hebrides, where rocks of the same character occur. Near Killochan, in the Girvan valley, is the "Baron's Stone," a huge boulder of granite nearly 40 tons in weight, which formed the "Hill of Justice" of the Lairds of Killochan. Its origin has been traced to the hills above Loch Doon.

Other peculiar surface features are the accumulations of stratified sand and gravel swelling into beautiful cones, called Kames. These are composed of material from the marine denudation of the boulder clay. Specimens of these occur near Skelmorlie, Maybole, and Muirkirk.

It must be apparent that the rock strata should have great effect upon the soil of a district, which in turn determines to a considerable extent the activities of many of its people. The basis of soil is weathered rock. The decay of most rocks gives us both sand and clay. Some approximate calculations have been given for the main soils of the county—seven parts moor and moss, six parts clay, and two parts sand or light soil. The reason why the first preponderates is due to the number of hills in the southern division and their extensive area along the eastern border. Clay or argillaceous earth is the next most common soil. It is found in all three divisions. In Carrick a range of it runs from a little south of Turnberry to near Kilkerran, where it crosses the Girvan and stretches toward the moors at the head of the parish of Kirkmichael. In Kyle it extends over almost the whole breadth of the district to within three miles of the shore. It stretches in Cunningham from the verges of the moors to within two or three miles of the sea, the parishes of Largs, West Kilbride, Ardrossan, and Kilbirnie being excepted. In places beds of it occur varying from 40 to 200 feet in depth, and it is put to commercially profitable purposes at the fire-clay and brick works in the neighbourhood of Kilmarnock, Hurlford, Dreghorn, Kilwinning, and Stevenston. A zone of light sandy soil, in some parts about a mile broad, runs along the shore from the Doon mouth to Ardrossan. Free from earthy matter it produces little vegetation except benty tufts. As advance is made from the shore the sand is more mixed with earth, and under the most approved methods of husbandry the soil has become highly productive. At Turnberry, Girvan, and Ballantrae the earth is of a similar character, and in other parts of Carrick it is gravelly. Some sandy soil occurs in the lower parts of the parishes of Dalmellington, New Cumnock, Galston, and in some parts of Craigie and Sorn. From the sea in Cunningham and stretching two or three miles inland, as well as in the lower tracts of West Kilbride and Kilbirnie and the greater part of Largs, the same description of soil is to be found. By a mixture of sand and clay in various proportions loams of different varieties occur according to the predominant element in each. In the holms along the river side and in other low situations the soil is of a loamy character. But the extent of this is small compared with what has been reclaimed by intelligent industry from large tracts of surface clay by the application of lime and other mineral manures.

7. Natural History.

It has been explained by the geologist that, at no great distance of time, as geological periods go, our islands were united to the continent of Europe. The English Channel, the Irish Sea, and the North Channel were at that time fertile plains, on the rich herbage of which animals we now call strange once browsed, to be pursued and devoured by others no less rare. In the same manner over these connected areas plants distributed themselves so far as their powers of migration and subsistence permitted.

Many creatures that once inhabited these islands have ceased to exist here. The skeletons of some of them have been found in various parts of the country. In some of our caves the remains of the mammoth, rhinoceros, tiger, lion, hyaena and wild horse have been unearthed. This circumstance denotes that for a time these animals attempted to naturalise themselves within our borders.

Again, it is worth observing that although the mainland of Europe has more kinds of birds and animals than we have in this country, both birds and animals are much more abundant here, and we have no quadruped or reptile, and only one bird, the red grouse, that is not to be found on the adjacent continent. The profusion of wild life in Britain compared with what prevails in other European countries is due to the fact that it is not usual here to shoot and trap small birds for food. Besides, our game laws, laws against trespass, and Wild Bird Act, have made of this country a vast sanctuary against the exterminator and disturber of wild life.

The character and natural features of Ayrshire are such as to make it rich in both animal and vegetable life. Its coast-line, where bold rocky headlands alternate with broad sandy beaches fringed by long dunes that guard the grassy "links," its rich alluvial plain with numerous water channels and moist shady glens, its wide stretches of moor and moss, its hills with rocky summits rising in some parts to considerable altitudes, its hedgerows and wooded tracts, all serve to make it a suitable home to a great variety of British *fauna* and *flora*.

Almost all the wild creatures native to Britain are represented by those now roaming this county. Rats and mice, shrews, voles, moles, and hedgehogs abound. Provided that food supply is easily attainable, rabbits are plentiful in the dry, shingly, sandy soil, and, in consequence, stoats and weasels are not far off. Hares are much scarcer. Squirrels are found in large numbers among the woods of pine and spruce firs, and foxes are numerous throughout the county. Bats emerging from their crannies are familiar all over as they flit with jerky movement in search of plunder. The otter is moderately common in suitable waters and among the rocks along the craggy shores. The pine marten is very rare, but one has been got at Maybole and another in Minnoch Water. The badger, too, is rare: Loudoun furnished

Guillemots nesting on Ailsa Craig

one specimen in 1894, and the Garnock a second as late as 1901. The beaver is now extinct, but remains of it have been found in the shell mound at Ardrossan as well as in Cleaves Cove on the Dusk. On our coasts the porpoise is familiar all the year round, while the common seal is a frequent visitor.

On and around Ailsa and along the rugged southern

shore there is no lack of life. Thousands of gannets crowd the dome-shaped islet, while vast colonies of gulls—common, black-backed, and kittiwake—guillemots, puffins, and razorbills make the neighbourhood throb with life. A few pairs of the herring-gull nest on the island, where the petrel is also not unknown. On the rocky shore south of Ballantrae the cormorant nests, and the shag, with a few black guillemots, is also found among the Carrick cliffs. Here, too, the chough and raven are frequent, and the blue rock-dove, the carrion crow, the hooded crow and the jackdaw are residents all the year round. The diver is no stranger to the rocky shore, off which the tern may also be seen to capture its prey, as it speeds with circling flight. The softer water edge is also alive with birds in search of their peculiar food. At the outlets of streams where food is abundant, or on heaps of sea-weed, dunlins, sandpipers, ringed plovers, with an occasional oyster-catcher, may be observed, and in autumn along the shore north of Girvan turnstones are numerous. The grey plover is a rare visitant, a specimen having been seen only once or twice at Lendalfoot.

Loch, stream, ditch, marsh, and the neighbouring leas are also lavish of life. The wild duck, goose, widgeon, and teal haunt some of the water expanses like Martnaham; the tufted duck is occasionally seen on Kilbirnie Loch; and the "coot ripples a long line across the reedy lakes." By a brook or swamp a gaunt heron settles, on the watch for fish or frog. Near by the snipe rises from the meadow where the pipit cheeps, and further off the whaup wheeples eerily, and the pee-wit utters in the stillness its well-known notes. From its perch over the stream a shy kingfisher from time to time darts down for plunder. In mid channel "the dipper keeps bobbing on a boulder," and on the sandy bank the wagtail with ceaseless activity searches for its insect food.

It would be superfluous to name here all the birds familiar to every schoolboy, whose habits and haunts in field or wood are the same in all parts of the country. Among the more common birds the starling is increasing rapidly in numbers. The magpie, on the other hand, is getting rarer in some districts. Linnets and larks are being considerably reduced in number, and with the decline of the flax-growing industry the goldfinch is moving southward. Red grouse, black grouse, and partridge are plentiful on the moors, and on the hills between Ayrshire and Kirkcudbrightshire is found the dotterel, "whose taking makes such sport, as no man more can wish." In sandy fields on the shores of the firth the corn bunting is common. The woodpigeon and the pheasant frequent the pine woods, the owl sits hooting on the spruce, and the elusive landrail creks in all the cornfields. The buzzard, the peregrine falcon, and the sparrow hawk are still not rare in the wild tracts of south Ayrshire.

Among occasional stragglers or annual migrants the chiff-chaff and the sedge-warbler are common summer visitors. The grasshopper warbler is occasional in some places, as at Beith. The pied flycatcher has been observed at Muirkirk, and the spotted flycatcher reaches remote and barren places like the head of Loch Doon. The brambling and the snow bunting are rare winter visitors, the latter being seen on Corscrine Hill south of Loch Doon, just over the Kirkcudbrightshire frontier. Some years ago flocks of parrot-crossbill were observed in Glen App and the Girvan Valley. The hoopoe, supposed to be the lapwing of Scripture, has been recognised at Coylton and Dean Castle. The hen harrier was formerly common in the wild tracts of the south, and the golden eagle was wont to bear its quarry to the higher peaks there. The white-tailed eagle bred on Ailsa, and the osprey nested on Loch Doon.

Many of the different kinds of British moths and butterflies have been seen in Ayrshire. It is recorded that even the peacock butterfly—called by one of our British naturalists the *regina omnium*—a rare type in Scotland, was once met with in the neighbourhood of Prestwick.

Of reptiles and amphibians the blindworm is sometimes found among dead wood and decayed leaves, among quarry refuse and in stone heaps. The adder is not uncommon in dry heathy places and sunny banks, and is sometimes seen in old ruins and under fallen trees. The crested or warty newt, the eft, the frog, and the toad are met with in the familiar places.

The mild moist climate of the west coast affords the very best conditions for the growth of ferns, mosses, and some kinds of fungi. Grey lichens line the bold front of rocks by the sea shore. They are also found on mountain summits and along the shores of Loch Doon. Among the Troon rocks there is an abundant supply of Irish moss.

The benty grounds along the sea shore furnish specimens peculiar to these places. Sea holly abounds, and the rare sea convolvulus is also there with thyme, heather, bluebells, dwarf roses, crowfoot, and the milk-wort. A local writer, Hewat, in his *A Little Scottish World*, adds the following—"The purple sea rocket may be found on the beach, as also the Isle of Man cabbage. The bird's foot vetch is found on the sandy soils, as also the highly aromatic chamomile, the delightfully scented true sweet briar or eglantine, the very rare figwort, the shepherd cress, and the spring vetch. In addition to the common corn poppy in the corn fields, there have been gathered the long prickly-headed poppy and the red poppy, and in the same places the corn cockle. There also grow the yellow as well as the white stonecrops. The fragrant yellow evening primrose has been gathered at Monkton, with the exception of Ayr, the only parish in the county in which it has been found. The blawort or blue bottle is also at times seen in the Prestwick corn fields, also the orange hawkweed in grass fields. The strong scented lettuce has been gathered near Adamton Mill, the only place at which it has been found in Ayrshire. Butcher's broom has also been got at Monkton."

Everywhere the landscape is ornamented with forest and other trees, natural and planted, the names of which one does not even require to know to appreciate their general effect upon the scenery. Where the soil is thin and poor the Scotch pine, larch, and silver pine predominate. The spruce fir with trunk erect, affording excellent cover for game, prefers the loamy soil at the hill foot. The ash, beech, birch, elm, oak, plane, Spanish chestnut, sycamore, and walnut grow best on good lands, neither too exposed nor too marshy. Specimens, numbering their years by centuries, are well distributed all over the county, a circumstance that shows that the climate is favourable to their fullest development. Of the oaks of Bargany, the beeches of Eglinton, the chestnuts at Loudoun, the firs of Cloncaird, and the limes at Montgreenan, some are magnificent. The winding woods of Fullarton contain noble specimens of pine, ash, sycamore, and elm. Close to Loudoun Castle there grows a venerable yew-tree, under which it is claimed one of the family charters was signed by William the Lion. One of the largest hollies in Scotland is at Fullarton House, where there is also a rare curiosity, an evergreen oak.

The algae or sea-weed along our shores is a form of plant life apt to be overlooked. Although these "flowers of the sea" are generally regarded as a *low* form of vegetable life, the structure of some of them is exquisitely beautiful. Their geographical distribution is well marked like that of plants on land, the temperature of the water being the main factor regulating it. Few places can show a better mixture of species than the Clyde sea area, where there is a great influx of southern forms due to the influence of the Atlantic Surface Drift. The sandy parts of the coast are very unproductive. Piles of sea-wrack are frequently seen there, but it is generally damaged and partly decomposed weed thrown up by the breakers after a storm. At the same time even there some rare specimens may be found growing in small tufts on the fronds and stems of large weeds. An abundance of species of beautiful form and colour may, at all times, be gathered along the rugged shores of the Ayrshire coast.

8. Along the Coast. General Features and Topography.

The coast-line of Ayrshire from the mouth of Kelly Burn in the north to that of Galloway Burn in the south, a distance of 84 miles, is of much interest and great variety.

At Kelly Burn and for several miles to the south beetling crags are observed to be close to the beach. About a mile from the northern extremity Skelmorlie is reached, a fashionable sea-side watering-place, sheltered from the east winds and open to the breezes from the west. Off the shore here is what is known to the Admiralty and the mercantile marine as the "Skelmorlie Measured Mile." It owes its popularity as a course over which the speed of vessels is tested, to the uniform and considerable depth of water near the shore as well as to immunity from troublesome winds owing to its comparative landlocked position. Southward extend the woods surrounding Skelmorlie Castle, an old baronial residence, picturesque and interesting. Precipitous hills are still close to the shore. About a mile and a half down the coast a conspicuous landmark, Knock Castle, overlooks the Clyde, a castellated dwelling of the Scotch baronial style. At the base of Knock Hill are the picturesque

The Bay, Largs

"Shamrock" and "White Heather," Fairlie-built yachts

woods of Quarter, and near by is Routenburn. In the richly wooded glen of the Noddle streamlet and about a mile from the shore is Brisbane House. Largs is soon reached on Largs Bay. The hills here recede about a mile from the shore permitting a "fine plot" to open to the view. Here the town is built in a position so sheltered that rare shrubs grow luxuriantly. The town, with the Gogo Burn flowing through it, though of ancient origin, presents the attractive appearance of a modern summer resort. After passing Kelburne House, the seat of the Earl of Glasgow, embosomed among its own wooded hills, the village of Fairlie is reached, pleasantly situated under the wooded slope which terminates the hills close to the coast-line. Here is a famous yacht-building yard, from which some of the fastest and fairest yachts of the R.N.Y.C. have been launched. Close to the sea on a rounded knoll near the brink of the deep and romantically wooded ravine of Fairlie Burn stands Fairlie Castle. A marked character of the coast here is the distance to which the tide recedes, leaving exposed a large expanse of mud. At the southern extremity of Fairlie Bay, just where the coast-line begins to curve eastward to form the Ardneil promontory, in a sheltered and beautiful locality is Hunterston House, the seat of one of the oldest families in the kingdom. After curving eastward the shore again trends southward, enclosing the lofty promontory of Ardneil Bank, supposed to be a raised sea-beach, which terminates in a great cliff a mile long, rising 300 feet sheer from the sea. This majestic perpendicular wall is buttressed in the south-west by Farland Head, or, as in

Fairlie Glen, near Largs

old prints, Goldberrie Head. Below the cliff and within the sea-wash on a ledge of rock are the ruins of Portincross Castle.

South of Farland Head lies Ardneil Bay, a beautiful crescent with musical sands, succeeded further down by others of a similar nature. The coast is flat, but at a short distance inland the land rises into a wooded platform, which ascends gradually southward until on a gentle acclivity fully half a mile from the beach stands West Kilbride.

The north bay of Ardrossan is now reached. It is fringed by a low fertile strip, from which there is an abrupt rise to a considerable acclivity formerly a sea-beach. The shore is sandy with low shelving rocks as Ardrossan is approached. Ardrossan stands partly on a rocky and rugged promontory extending south-west about half a mile into the sea. The pretty sea-bathing bay further south, known as the South Beach, is fringed with many fine mansions, and ends at the town of Saltcoats.

The coast now takes a graceful sweep for about ten miles to Troon, enclosing what is called Irvine or Troon Bay with a stretch of beautiful sands over most of that extent. Immediately south-east of Saltcoats the sands of Stevenston are passed, and beyond that a long spit of land between the sea and the estuary of the Garnock is reached. This expanse is occupied with the Ardeer sandhills, a vast accumulation of drift-sand. Irvine Harbour, formed by the combined estuaries of the Garnock and the Irvine, is now passed, and for several miles beyond to the north sands of Troon the coast remains absolutely flat, and the

Ardneil Bank, West Kilbride

bent lands adjoining the shore are mostly occupied with a succession of golf courses. At Shewalton, south-east of Irvine and about two and a half miles from the shore, there is another raised sea-beach with quantities of shells and other organic remains discovered on it. The adjoining lands of Gailes and Barassie form a vast plain extending from the base of Dundonald heights north-west to the sea. Troon, a sea-port and favourite watering-place, stands on a promontory extending over half a mile into the sea. Curving to the right it forms a picturesque natural harbour sheltered from nearly every wind. The South Sands, circular in form and well adapted for sea-bathing, form the northern extremity of Ayr Bay, which extends in a big curve for some miles to the rugged cliffs forming the escarpment of Brown Carrick Hill. For almost five miles beyond Troon the shore is of the same character as that just described, with sandy bluffs here and there forming a natural low wall against the encroachment of the sea. The bents that border the shore are occupied with the famous golf-courses of Troon and Prestwick. About midway between these places last named stands Monkton, where six roads meet. Prestwick, an old burgh of barony, is an attractive watering-place and popular health resort, famous for its mild climate and magnificent sands. Two and a half miles beyond Prestwick the busy harbour of Ayr is reached, and then a delightful stretch of sand with no definite line between the beach and the Low Green of Ayr extends for two miles to the mouth of the Doon. Immediately beyond the Doon the shore abruptly bends to the west and gradually rises into precipitous steeps broken with many creeks and caverns, and extending to the Heads of Ayr, the rugged foreland of Brown Carrick Hill—a landmark of some importance. A prominent object on this shore is Greenan Castle, perched on a rocky ledge overhanging the sea. The coast, of the same bold character, again curves southward to Dunure Point. Dunure village is close by, with its little fishing harbour and the ruins of Dunure Castle. At the other end of Culzean Bay, five miles further south, is Culzean Castle. The rocky height on which the castle stands proudly dominating the sea seems to dwarf the stately dimensions of the pile. A remarkable feature is the number of caves and galleries that penetrate the hard rock.

The shore forming the southern extremity of Culzean Bay, fringed with trees, again bends south-west to the wooded bluff of Bawin Point. Round this point the little fishing village of Maidens is soon reached, a comfortable retreat much in favour as a summer residence. The village owes its name to two islets lying in the bay. A mile south-west is observed a rocky promontory culminating in a lighthouse and the ruins of the old Scottish keep of Turnberry Castle. Among the dunes and along the undulating sward adjoining are the famous Turnberry golf-links, with gently sloping back-ground of field and wooded hill, surmounted by a palatial hotel. At the southern extremity of Turnberry Bay and close inshore lies Brest Rock, which had a bad reputation for shipwrecks before the erection of Turnberry lighthouse. A mile and a half further south there stood at one time a chapel dedicated to St Donan, the meagre remains of which, including the font, may still be seen. Girvan is soon after reached, a much frequented sea-side resort with many attractions. The feature of this part of the coast is the pebbly beach with stretches of sand outside.

South of Girvan Shallock rocks are passed, and further on the castle of Ardmillan. The coach-road which

Kennedy's Pass, near Girvan

borders the shore for fourteen miles assumes here a kind of switch-back character, rising and falling as the contour of the ground necessitates. At Kennedy's Pass, a picturesque defile on the way, the scenery is very striking. Towering cliffs against which the angry sea has chafed for ages, streaked and spotted with lichen, overhang the way, and a talus of huge fragments from the impending cliffs like the débris of gigantic quarries lines the shore. Over more than nine miles of coast here there is a marvellous field for the geologist.

Lendal, a tiny hamlet at the foot of Lendal Water, six and a half miles from Girvan, lies in a quiet bay where many a craft with contraband cargo used to seek shelter. Perched on the edge of a gully is Carleton Castle. At the base of Carleton Crag a cluster of cottages called the Fishery is reached. After passing Chapman's Crags there begins the long, steep, picturesque ascent of Bennane Hill. Near the top are the ruins of the Craignaw Inn, associated with tales of smuggling days. Here the scenery, grand and impressive, has evoked the admiration of men like Murchison, who has proclaimed these cliffs to form a scene of igneous rock as pictorial as may be found in any other part of Scotland. Penetrating deep into the foot of the hill is Bennane Cove, identified with a scene of barbarity in Crockett's *Grey Man.*

After rounding Bennane Head Ballantrae appears with fine effect backed by green hills. Formerly a notorious smuggling centre, the town has long been famous for herring and cod fishing. It is also a popular sea-side resort.

South of Ballantrae, round Downan Point, the scenery is of the same awe-inspiring character as that which obtains at Bennane Head. Cove and cliff, the haunt of rare birds, alternate till the mouth of App Water is passed, and a mile further south along a pleasant shore the southern extremity of the county coast is reached, where the Galloway Burn discharges into the Channel.

9. Coastal Gains and Losses. Sand-banks. Lighthouses.

Many evidences go to show that some remarkable changes have taken place along our coasts. At various parts some distance back from the present shore line we observe an old sea-platform raised high and dry beyond the reach of the waves. We may even find a succession of these terraces at a distance from each other and at varying heights. The explanation is that the land has been elevated and that these successive platforms were the work of the waves during the successive pauses in the upheaval. Additional proofs of this land movement are to be had in the fossils of the geological strata, and in the numerous sea-caverns along our cliff-girt shores, now several feet above the highest spring tides. While these are given as evidences of land upheaval there are no less conclusive proofs of the opposite movement, land subsidence. Submerged forests are not uncommon round some parts of our coasts where the land has slowly settled under favourable circumstances. The black peaty matter cast up by the sea after a more than usually violent gale consists of dark clay filled with pieces of moss washed up from beds submerged below the ocean floor. From the nature of things this vegetable substance must have been formed on dry land. These phenomena simply show that the changes have not been constantly in the same direction over the same area. Although there is little evidence on our coasts that such movements are in process, we have some indications that they are at present going on slowly. The greatest changes on our coast line, however, are not due to any elevation or subsidence, but to the rapid and constant devouring of the shore by ocean billows carving the coasts into narrow creeks or deep bays, and fringing them with rocky islets or shingly deposits.

On an examination of the sea-line of Ayrshire it is found that many of the changes mentioned have actually taken place and that others are in operation. Raised beaches have been discovered at varying altitudes throughout the county, each representing a separate epoch. The last epoch is indicated by a terrace about 10 feet high, an earlier one has a platform 40 feet high, while the cliff at Portincross, 300 feet high, represents one still earlier, and near the "Windy Wizzen" east of Darvel an old sea-beach 700 feet above the present sea-level is pointed out.

The character of the coast-line is determined by that of the general surface. The bold front presented by the northern and southern parts of the county is due to the hard and uniform texture of the rocks and their ability to resist wasting influences. The large indentation in the middle forming the bays of Troon and Ayr is mainly caused by the exposure of softer material to the influence of prevailing winds and marine currents operating over a lengthened period. The sweep of many miles of unbroken Atlantic swell from the south-west, acting on a sandy shore for ages, must have been of great force in modifying the land, and in cutting it out to its present shape.

If the esplanades constructed at considerable expense

by some of the more enterprising coast towns are disregarded, there are no groins properly so called to prevent sea-encroachment. To protect that part of the coast most liable to marine erosion, nature herself has constructed a sandy ridge, held in place by bent and tufty grass and extending just where its usefulness is most necessary.

A broad submarine flat slopes gradually westward, from which here and there project rocks of exactly the same appearance and height as those of the mainland adjoining. Within recent times the shore between Fairlie and Farland Head must have undergone changes if Pont, a topographer of the seventeenth century, was not deceived by the appearance of the beach at low tide. In his work on Cunningham he alludes to a small island adjoining Fairlie harbour, which he calls *Fairlieland*. This island, if it ever existed, has disappeared. As has been already mentioned, the shore here at low water exposes a vast expanse of mud, which at certain conditions of the tides presents the appearance of an island. It is probable that this circumstance may have given to the topographer a wrong impression.

North-west of the mouth of Ardrossan harbour and forming a natural breakwater to it is Horse Isle, so named from Philippe Horsse, son-in-law to Sir Hugh de Morville, reputed founder of Kilwinning Abbey and Lord of Cunningham. It contains nearly 12 acres of good pasture with abundance of fresh water. Of no great height above the level of the sea, it has a beacon tower erected on it.

In the bay of Troon three miles from its southern extremity and one and a half miles from the shore are the Lappock Rocks, and, lying off the north sands of Troon, is Mill Rock, a low ledge, on which herds of seals are occasionally seen.

In the south sands a number of rocks and reefs occur. At no great distance from the beach lies Seal Island, discovered only at low tide. Seals visit it only on rare occasions. Further south a quarter of a mile from the shore is Black Rock, of bald trap without a green blade, stretching half a mile into the sea. It is never quite flooded except when the spring tides are accompanied with a south-west gale. Exactly three miles off the South Beach of Troon is Lady Isle, about a mile in circumference. It is never covered by the tide, and has, therefore, acquired vegetation enough to feed a swarm of rabbits. Two beacons have been erected on it, one larger than the other, intended not only to ward vessels off the rocks, but when seen in line to guide them to safe anchorage in Troon harbour. These islets lying in a portion of the Firth where it has made considerable encroachments on the land would point to a time when they all formed part of the mainland.

In Maidens Bay are two islets, and south of Turnberry is a reef called the Brest Rocks with Balkenna Isle. On the former a beacon has been erected.

In lone sublimity and with graceful symmetry nine and a half miles west of Girvan stands Ailsa Craig, the sentinel of the Clyde, welcoming and taking leave of thousands of vessels belonging to all quarters of the globe.

Ailsa Craig

It is a great conical rock about two miles in circumference rising abruptly out of the sea to a height of 1114 feet. The immensity and grandeur of it can be ascertained only by sailing round it or by scaling it. It is accessible only on one side. The rock is a hard granite assuming in some parts a distinctly columnar form. On the north-west the columns present an unbroken elevation of nearly 400 feet—a height greater than has been elsewhere found among rocks of a similar formation. On the other side the craig descends to the sea with a steep slope terminating in a small beach formed of the débris which in course of ages has been detached from the vast mass. The particular kind of granite found here is honoured by the distinctive name of *Ailsite*, which, being very hard and close grained, takes on a beautiful polish, and, consequently, makes an ideal curling-stone. A new industry started here is the quarrying of granite setts, which are shipped to all quarters of the British Isles. It is the home of a few goats and many rabbits, and eight different kinds of sea-fowl breed on the cliffs in countless numbers. About 200 feet from the summit are several springs and a pool, round which grows some rank herbage ; and on the ledge of a crag on the eastern front are the remains of a chapel and an ancient sixteenth century stronghold of three storeys. In later times the rock became a notorious smuggler's haunt.

To assist the mariner in navigating our river-mouths and shores, fog-signals, beacons, buoys, and lights are provided at various points. The beacons have been already mentioned, and there are minor lights connected with the harbours of Ardrossan, Irvine, Troon, Ayr, and Girvan. On Lady Isle, besides the beacons, there is a buoy lit with gas, which works automatically. The only important lighthouses along the Ayrshire coast are those of Ailsa and Turnberry. In 1883–86 a lighthouse and two fog-signals were erected on Ailsa. The light is a six-flash one of 15 seconds' duration, with an interval of

Turnberry Lighthouse and Castle

darkness. At the northern and southern extremities are the two fog-sirens whose weird sounds can be heard 20 miles away. The lighthouse at Turnberry crowns the promontory which projects seawards from the long stretch of sand-dunes there. The light is both revolving and intermittent.

10. Climate and Rainfall.

Climate is the general tendency of a country to certain weather conditions with respect to temperature, moisture, sunshine, and wind. Weather is the periodic departure from or return to any of these conditions. Climate holds the same relation to weather as a man's constitution does to his health. It has been defined as average weather, and depends upon various circumstances—especially upon the latitude of the district, its altitude, its nearness to the sea, its soil and vegetation, its slope, the prevailing winds and the ocean currents near.

Before encountering some statistics bearing on the atmospheric phenomena it may be well to note that these are of value in so far as they are put to practical use in applied meteorology. The prosperity of any particular district and the pleasantness of its climate are largely due to temperature and moisture, which in turn depend not only on the amount of sun and rain, but on the relative distribution of rainy or sunny days. In addition to these it is well known that sunlight, and the direction and force of winds at certain seasons, are of great importance to plant life. The losses of crops due to insects are sometimes enormous, and these depend to some extent on meteorological conditions of air and ground. Again, vegetation is frequently liable to disease. This is due to climatic influence. Various degrees of moisture, heat, and sunlight are favourable to the development of fungus, which is due to the growth of vegetable parasites.

The average weather conditions of the British Islands are greatly influenced by the prevailing south-west winds, which come warm and moist from the Atlantic. The Firth of Clyde, both from its shape and its situation, is admirably suited to get full advantage from these conditions, and to communicate in turn its benefits to Ayrshire.

The predominant winds in Ayrshire are those from the south, south-west, and west. Those from the south-west blow for more than half the year.

Owing to the direction of the prevailing winds rains in Ayrshire and in the south-west of Scotland as a whole are frequent, often copious, and sometimes of long duration. This is due to the comparatively warm air blowing over an immense surface of water and bringing with it a large amount of aqueous vapour. On reaching land, especially if there is a rapid rise in its surface, as in the south-west and north-west of the county, the air is forced to move upwards as well as forwards. It is consequently cooled by expansion, and, its capacity for retaining water vapour being diminished, it gets rid of the vapour as rain. Fogs, too, and thin drizzly rain are frequent, especially in spring and autumn. These are due to the warm moist air from the sea coming in contact with the colder clay soil and moss earth of which there is such a preponderance in the county. From the higher grounds in the early mornings the mist often appears like a sea of fleecy vapour with the loftier uplands lying like islands on its bosom. It generally disappears with the advance of day. Some years snow begins to fall about the middle of November

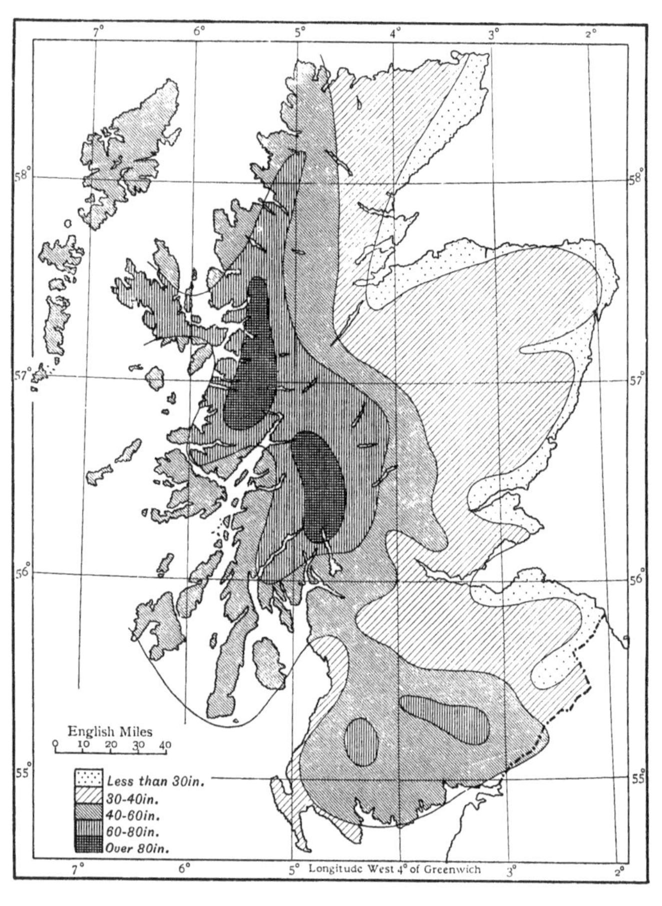

Rainfall Map of Scotland. (After Dr H. R. Mill)

and continues its periodical visits till March or even April.

In the year 1908 observations from eight stations in Ayrshire gave an average rainfall of 49·07 inches, the highest being 77·8 inches at Shaws Waterworks, Kelly Dam, the lowest 29·89 inches on Ailsa Craig. The lowest measured on the mainland was at Mauchline, 33·78 inches. In the year given the highest rainfall in Scotland was 123·7 inches at Achnashalloch in Inverness-shire, and the lowest 19·13 inches at Dalkeith Gardens. It comes to this—that if all the rain that fell that year had stayed on the ground, and had not run away or sunk into the earth, the water over Ayrshire would have been 49·07 inches deep at the end of the twelve months. When it is considered that an inch of rain weighs fully 100 tons over an acre of ground the effect of the rainfall over the county can be realized. During the same year rain fell at Dalry on 228 days, at Dalrymple on 222 days, and on Ailsa Craig on 158 days. That year appears to have been a particularly wet one, for during a period of 30 years (1870–1899) the approximate average over the county was only 46·67 inches. This shows an increase for that of 1908 of 2·4 inches, or of 5·1 per cent. over that term of years.

The sunniest district in the British Islands will naturally be along the shores of the English Channel. But even the most favoured places get no more than 1800 hours bright sunshine throughout the year. The average duration of sunshine in Ayrshire is between 1200 and 1300 hours a year, or approximately 3½ hours a day.

As the sun is above the horizon in Britain for about 4400 hours in a year, Ayrshire is favoured with about 28 per cent. of possible sunshine. The sunniest months in the year are generally May, June, and August. In 1908 these months had respectively 180, 196, and 186 hours of bright sunshine. The gloomiest month was that of December, which had only 30 hours of sunshine during its 31 days.

11. People—Race, Type, Language, Dialect, Settlements, Population.

In the earliest historical accounts of the races Strathclyde, of which Ayrshire was a part, is said to have been in the possession of a powerful Celtic tribe called the Damnonii, who are supposed to have come from the same stock as the Celts in Cornwall. The investigation of tumuli, in the chambered graves of which skeletons with pronounced Celtic cranial characteristics were discovered in stone coffins and with stone implements, helps to fix approximately the time of their first appearance. After the withdrawal of the Roman legions at the end of the fourth century the Picts from north of the Clyde and the Scots from the west invaded Strathclyde, colonising the northern parts. Britons, Picts and Scots were all Celtic in speech, and this is clearly seen in the place names of the shire.

Angles also crossed the hills from Northumbria. After 1066 refugees trooped north to escape the Conqueror, and in the twelfth century many Anglo-Normans came, as the de Morvilles and the Fitzalans.

In later times Ireland and our own Scottish Highlands have contributed not a little to the leavening of the people. Naturally in the south of the county there is a greater admixture of outside elements than in the northern parts. Along the highway from Ayr to Portpatrick at the beginning of the nineteenth century there were constant streams of Irish beggars passing through, some of whom, like the débris of a moving glacier, would fall off by the way and get located in the district. Before the introduction of machinery into the harvest field, huge bands of Irish and Highlanders made their way into the county each year to assist in the autumn operations, and the practice is still being continued in connection with the early potato crop. Railways, too, have induced a migration of the general public such as was not known before the country was opened up by railway enterprise. Several industries in the county have also brought a not inconsiderable influx of aliens within her borders, notably French, Italians, and Russian Poles.

The inhabitants of the county being of such a cosmopolitan character, few are to be met with of a pure type with the peculiar features of the original race. We are told that of all the men of Great Britain those of south-western Scotland are distinguished for their tall stature, which is here of greater average than that attained in any other district of the British Islands. But, owing to remote and later fusions, we fail to discover here more than in any other part of the Lowlands the prominent cheek, the bright eye, the more or less light hair, the regular features, and other factors relied on in a scheme of classification which are said formerly to have characterized the inhabitants of Ayrshire.

The Scottish language is now the name applied to the Teutonic speech of the lowlands of Scotland, which came from the northern dialect of Old English. For long, the Scottish language meant the language of the Scots of Ireland. It was called *Erse* from the Scotch form of the adjective *Irish* (*Ersche* or *Yrische*) to distinguish it from *Inglis*, by which the language of the early Scottish writers was known. Walter Kennedy, born about 1450 and son of the first Lord Kennedy, is called by the poet Dunbar an "Ersch Katherane" in reference to his extraction from the Celts of Carrick, and in reply says to Dunbar—

> "Thou luvis nane *Erische*, elf, I undirstand."

This was the language still spoken by many of the inhabitants of Ayrshire in Kennedy's days, and it is said to have been used at Barr as late as the end of the eighteenth century. But latterly in Ayrshire as in the Lowlands generally a form of northern English became the vernacular.

In voice and pronunciation Ayrshire speech has some marked features. Round Kilmarnock and in the northern district generally the Scotch guttural prevails, as in *loch*, and is often made to do service for *p*, *b*, *t*, *d*, *k*, *th*. About Old Cumnock the voice gets less guttural, becoming more nasal as New Cumnock and the county frontier is reached, where it approaches more to the English tone. In Carrick the guttural is also less in evidence, the dental being more pronounced throughout this district. The consonants *c*, *p*, *t*, *f*, *s* are subject to modification and made to sound like *g*, *b*, *d*, *v*, *t*. The sound of *th* changes to *d*, and *with* is pronounced *wid*. This modification, along with the addition of *t* to *once* and *twice*, which is very general, is no doubt a relic of Irish influence.

The question of vocabulary is also an important one, and we have merely to glance at a glossary of Burns's works, or scan the homely portraiture of Galt with its rare mastery of the Scottish dialect, to come upon many words whose circulation does not extend far beyond the borders of the county.

The population of Ayrshire according to the census of 1901 is 254,468. A hundred years previous the population was 84,297, or about one-third of what it is at present. During the same period Lanarkshire increased the number of its inhabitants ten times. There are about three acres of land in the county to every man, woman, and child, or 221 persons to the square mile, as compared with 937 in Renfrewshire, and 64 in Wigtownshire. Sutherland, the most sparsely inhabited county, has nine people to the square mile, while Lanarkshire has 1524. The average for the whole of Scotland is 151, and for England and Wales 558.

12. Agriculture—Arable, Pastoral, Woodland Areas, Main Cultivations, Stock.

Husbandry in one or other of its forms, tillage, dairying, or sheep-rearing, is one of the main industries of Ayrshire. Throughout the county there is no lack of agricultural enterprise in spite of climate or weather risks which can neither be disregarded nor avoided. Grain-cropping, especially that of oats, is well distributed over the county; green-cropping is more general near the coast; dairying in all its branches is well represented in the north and north-east, as well as in some parts of the Ayr basin; sheep-farming is mainly to be found in the eastern uplands and in the hilly tracts of Carrick.

Ayrshire covers an area of 735,262 acres. Of this in round numbers there are 47,000 acres in corn crops, 18,000 in green crops, 255,000 in hay, clover, sainfoin, grasses and permanent pasture (the last covering 167,000 acres), 327,000 in mountain and heathland used for grazing, 27,000 in woods and plantation. What remains is under water and bog.

Of the corn lands 45,000 acres are in oats, the remainder being under barley or bere, wheat, rye, beans, and peas. Potatoes account for 10,000 acres of the green crop, turnips and swedes for 6000 more, while mangolds, carrots, cabbages, and other minor products cover 2000 more. From these figures it will be observed that oats form the main corn crop in the county. Rye is seldom sown except in small quantities on sandy soil near the coast. Wheat is gaining in favour and yielding valuable returns on the richer and heavier soils. Green crop husbandry, especially that of the potato and the turnip, has entirely changed the whole system of cultivation, and a liberal use of lime, guano, and other auxiliary manures chemically combined with the soil, has enormously increased the amount of agricultural produce. Early potatoes are now raised in great abundance, and the districts round Girvan and West Kilbride, with special advantages of soil and climate for this kind of cropping, are identified with their scientific cultivation. Energy and enlightened enterprise call forth all the fertility of the fields, and immense quantities are produced which find a ready and profitable market in the towns of Scotland and the north of England. Carrots and mangolds are grown more extensively in Ayrshire than in any other Scottish county.

A great part of the pasture is occupied with dairy stock and cattle feeding. The dairy forms an important element in Ayrshire farm management. The value of the milk and butter produced is very considerable and the far-famed Dunlop cheese is a valuable output. The commercial value of this cheese, however, did not compare well with some English varieties. The consequence was that about the middle of last century the County Agricultural Association along with some enterprising proprietors took measures to have the Cheddar method introduced. This has been attended with great success both in Ayrshire and Galloway, where the women are widely famed as cheesemakers; and the annual cheese show in Kilmarnock is now the most important in the kingdom. This great change has brought wealth to the county. It is reckoned to have increased the annual value of a cow's product by about £2 and there are 51,000 cows in Ayrshire. On the milk-producing lands of the north it is the usual practice for the farmer's household to be astir at 3 o'clock in the morning to have the milk forward for the Glasgow breakfast. This habit may soon be somewhat modified, as local creameries are being established here and there, and these contract with surrounding farms for their milk supplies.

System has been put into the breeding of cattle with the result that the Ayrshire shorthorn is much prized for the purposes of the dairy as well as of the butcher. On some of the farms round Ochiltree and in the Irvine valley the breeding of this animal has reached a high standard of perfection. The improved Ayrshire cow is distinguished by certain characteristic features—effeminate shape, heavy hind-quarters with large udder, small fore-quarters and neck, an intelligent twinkle of the eye peculiar to what is said to be the most highly improved breed of cows in the world. It varies in weight from 20 to 40 stones. The quality for which it is most valued is in the abundance of milk it yields. Ten Scots pints (1 Scots pint = 3·00065 imperial pints) per day is not uncommon; some give 12 or 13; and even 14 is not unknown.

A large proportion of the county, four-ninths of the whole, is covered with mountain and heathland suitable for sheep-grazing. Flocks thrive steadily on high lying grounds where the soil rests on beds of limestone or Silurian rocks, the grass being fine, short, and thin, but sweet and wholesome. Sheep that feed along the dank bottom of a valley get plump and fleshy, but at the fall of the year braxy attacks the flock, and the result of the pestilence more than neutralizes to the farmer the apparent

Ayrshire Cow

advantage gained from the fertility of the soil. At local lamb sales the size and fleshiness of the animals generally attract the fancy of the novice. These factors, however, do not regulate the price, which is determined mainly by the place where the lambs have been bred. It is not uninteresting to know that the life of a sheep is worth fewer years' purchase when it is reared on the luxuriant haugh than on the elevated hill-top where the conditions are prophylactic and tonic.

During the last century and a half great changes and improvements have come upon the agricultural conditions of the county. Much of what appeared to be unreclaimable wilderness in the form of arid moor or quaking bog, has been brought under cultivation and is now rich fertile land. Self-willed clay soil was a cause of much trouble to the agriculturist. Retentive of moisture it could be ploughed only at suitable times, it refused to be much impressed by the harrow, the seeds were scarcely covered, and the roots would with difficulty penetrate the tenacious clods. With the employment of new and improved implements, and under an enlightened course of husbandry, these difficulties are now rarely felt. Even early in the seventeenth century the northern parishes, where there was a deep fat clayey soil, yielded a great deal of excellent butter. The pastures were enriched with lime and this practice is said to have caused one acre to yield more than three acres in any of the next adjacent counties. In the same district, especially in the neighbourhood of Beith, large quantities of flax were raised, but at the beginning of last century linen largely ceased as an article of merchandise in the town and the fields were put to a more profitable purpose.

It is proved by authentic history that large tracts of the county were at one time covered with dense natural forests, and there is additional evidence in the trees found in the mosses, in remnants of birch and mountain ash still growing in patches here and there, and in the place-names in different parts of the county. Fordun says that the forest of Selkirk extended as far west as the Castle of Ayr, and it is well known that almost all the district of Kyle was covered with wood up to a very recent period. It appears from a charter granted to the monks of Melrose that the parishes of Muirkirk and Sorn were forest at the end of the twelfth century, and at the time of the Reformation woods extended from Ayr for upwards of 10 miles in the direction of Barnweil, the church of Craigie being called *the Kirk of the Forest.* Besides, it is related that the monks of Crossraguel kept themselves busy developing the resources of the extensive forest surrounding their monastery, and the Cistercian monks from the monastery of Mauchline are said to have pastured part of the great forest of Kyle. Laglain Wood lying north of Coylton, to which the patriot Wallace had often to retire at the commencement of the War of Independence, is said to have formed part of the same natural forest.

Few natural woods of any extent now exist anywhere in the county. Stray remnants may still be seen dragging out a precarious existence on weather-beaten braes, or by narrow ravines "like battered sentinels on a forlorn field," and patches may here and there be found intermixed with modern plantation.

The place-names testify to the former existence of considerable tracts of woodland. *Dalblair* in Kyle is a dale cleared of trees. *Shaw* occurs in several parts of Ayrshire, as Round*shaw*, Weit*shaw*, and Lain*shaw*. Names in *-wood*, are still common in every parish, while there is a sprinkling of names in *den* (a deep wooded valley) and *hirst* (a thick wood). These last prove not only that trees had grown in these situations but also that woods were in existence after the Saxon language began to prevail over the original tongue, probably about the thirteenth or fourteenth century. In the neighbourhood of Beith, which itself is a Celtic word meaning *birch-tree*, there are places called Rough*wood*, Threep*wood*, Full*wood*head, *Wood*side, Hazel*wood* (now Hazelhead or Hessilhead), and in localities a little further remote there is a Nettle*hirst*, a *Den*, a *Blair* (a plain clear of wood), all indicating that the district at one time must have been extensively leafy.

Arboriculture has greatly advanced, and numerous small woods now abound in the basins of the rivers where they exist for embellishing the landscape, or for the purpose of giving cover for game and shelter to the fields.

As we should expect in a district so largely given to husbandry the numbers of the various kinds of live-stock are considerable. They are naturally always varying in number, but, approximately, there are 10,000 horses, 100,000 cattle (51,000 cows), 380,000 sheep, 14,000 pigs. The relative numbers in four other counties are :—

	Horses	Cattle	Sheep	Pigs
Dumfries	8000	65,000	580,000	9000
Renfrew	3500	26,000	44,000	1300
Perth	13,000	72,000	690,000	8000
Edinburgh	4500	19,000	187,000	9000

13. Industries and Manufactures.

The manufactures of Ayrshire have attained considerable importance. After the middle of the eighteenth century the spirit of enterprise began to appear over the county. At first water-power, everywhere ample, was utilized, and a new era began with Watt's improvements on the steam-engine. By and by, when more ready inter-communication was established by means of the railway industry, manufacturers even in smaller places were able to compete with those nearer the market on terms not altogether unequal. For although freight added to the cost, the additional expenditure was counterbalanced by lower rent, taxes, and wages. This will serve to explain why so many comparatively small places in the county are still able to compete successfully in the industrial markets.

At the capital of the county the industries are varied, some of them extensive. Factories for carpets, lace, woollens and winceys have been established here, as well as works for tanning, currying, and shoemaking. There are also ship-building, coach-building, iron-founding, sail-making, and rope-spinning industries with prosperous woodyards, sawmills, chemical works, and dye-works. The manufacture of carpets was introduced into Kilmarnock in the year 1777, where it is still carried on. The printing of calicoes was begun in the same place in 1770 and of shawls in 1824. There are also here several extensive tanneries besides factories for the manufacture

Lace Machine (Morton's of Darvel)

of shoes for exportation, tweeds, winceys, and cotton goods. Woollen and worsted spinning is also carried on. Owing, however, to the situation of the town in the midst of a great mineral district the iron trade may be said to be the staple industry; and several large machine-making, iron-founding, and hydraulic engineering establishments have been set up. In 1858 the Glasgow and South Western Railway Company transferred to Kilmarnock their large engineering shops, which supply nearly the whole of their system with locomotives, carriages, and waggons. Fire-clay goods and sanitary ware are extensively made in the vicinity. At Irvine, too, the trade is of a varied character. Here, there is considerable activity in ship-building, coach-building, engineering, and metal-casting. But the prosperity of the town is due in no small degree to the impulse received from the establishment of two extensive chemical works for the manufacture of bichromate of potash and alkali.

In the smaller towns the manufacture of textile fabrics may be considered the principal. At Catrine and at Patna there are extensive cotton works. Woollen manufactures in various forms are to be found at Stewarton, Dalry, and Old Cumnock; and along the valley of the Irvine, notably in the towns of Galston, Newmilns, and Darvel, there are flourishing factories for lace, muslin, and tapestry.

In Ayrshire the manufacture of linen was formerly much more extensive than it is now, and farmers used to find it profitable to raise considerable quantities of flax. It is still an article of manufacture at Barrmill and

Nobel's Works, Ardeer

Kilbirnie, where by an industrial inertia the mills remain although they are now entirely dependent on imported raw material.

Extensive cabinet and chair-making works are established at Beith; and stretching about three miles among the sand-hills on the coast near Stevenston an important

Glengarnock Ironworks: Converter blowing

dynamite factory is located. Ayrshire embroidery and needlework, consisting of beautifully executed patterns in muslin and cambric, for which the district was long famous, are still produced to some extent in different villages of the county. Immense fields of ironstone have been opened up in Ayrshire and the iron trade has risen to great importance. There are works at Muirkirk,

Glengarnock, Stevenston, Eglinton (Kilwinning), Hurlford, Lugar, and Glenlogan. In all, there are about 31 furnaces in blast, producing annually 450,000 tons of pig-iron.

The production of salt by evaporation from sea-water was formerly prosecuted at several places along the coast,

Glengarnock Ironworks: Soaking pits

especially at Prestwick, Troon, and Saltcoats. The industry declined in consequence of the repeal of the salt-tax and the production of finer and cheaper salt in Cheshire and Worcestershire. The town of *Saltcoats* is the only one in the county deriving its name from the industry associated with it. The term implies the *cots* occupied by the *salt*-makers. It may be added that, like the colliers, the salters in Scotland were, from the beginning of the seventeenth to the end of the eighteenth century, in a state of serfdom. They were bound to the salt-works; and when the property was sold, they were transferred with it to the new owner.

The excellent roads and railways by which its industrial products can be conveyed readily and cheaply to the centres of distribution give Ayrshire a great advantage in its trade facilities.

14. Mines and Minerals.

The more important minerals raised at the present day are coal, ironstone, limestone, sandstone, and clay. The coal amounts in round numbers to 4,140,000 tons yearly, iron ore 319,000 tons, limestone 18,000 tons, sandstone 104,000 tons, fire-clay 165,000 tons, brick-clay 8000 tons.

The coalfields are of great extent and mainly lie in the two northern divisions of the county. The only coalfield in Carrick is that in the Girvan valley mostly in the parish of Dailly, "a little piece of coalfield separated and jammed between the hills," as Geikie described it. Coal was worked here from very early times. At Drumochrin north-west of Dailly it was mined as far back as 1617. There is a large and rich coal basin in the south-east of the county at New Cumnock, where almost every coal seam represented in Scotland is to be found, their aggregate thickness being about 80 feet.

Those of the lower section are now mostly worked out, together with some of the fine seams in the upper section. But there is still a great store of splendid coal. It is calculated that more coal remains to be worked in Old Cumnock than in any other of the adjacent parishes. There are also extensive and rich basins of it in the parishes of Muirkirk, Riccarton, Kilmarnock, Kilmaurs, Dreghorn, Kilwinning, Stevenston, and Dalry. At Stevenston the measures continue to a considerable extent under the sea. Coal is also successfully worked though to a less extent in the west half of Auchinleck as well as in the vicinity of Beith, Kilbirnie, and Stewarton. There are some valuable seams considerably broken with trap dykes in the western part of the parish of Galston; and under the whole parish of Tarbolton except in the south-west they lie at too great a depth to be profitably worked meantime. A large part of the coal is sent across the North Channel to Belfast and other manufacturing towns in the north of Ireland.

The production of iron-ore is now greater in Ayrshire than in any other Scottish county. Lanarkshire comes second with an annual output of 277,000 tons. Iron is generally found in the same basin as the coal, but the most valuable seams occur in the neighbourhood of Dalmellington and Muirkirk. Veins of haematite, as valuable as they are rare in the county, are found in Muirkirk and Sorn, and also to a small extent in Carrick.

The usual method of mining is practised in Ayrshire, that is by shafting and raising the material to the surface by means of engine, rope, and cage. But wherever the contour of the ground allowed, adit-levels or tunnels, locally known as "inganees," were formerly employed to some extent. So far as is known the only one in use at present is in the neighbourhood of Cumnock.

Limestone is also widely distributed over the coalfields and is abundant both with and without fossil remains. Large quantities richly charged with marine organisms are obtained in the neighbourhood of Beith and New Cumnock. In the former district it is not now quarried and burnt to the same extent as it was twenty years ago. But there are numerous remains of workings and kilns.

Excellent sandstone for building is found throughout the county. Near Mauchline is the great Ballochmyle quarry of red free-stone, a building material which suggests both warmth and richness. There are two quarries of fine sandstone on the side of Dundonald Hill, and from them some very perfect specimens of fossil plants are obtained. Some trap pierces the sandstone, which hardens it and makes it suitable for carving. One of the most valuable sandstone quarries in Scotland is on the estate of Ardeer near Stevenston. Beds of building stone are also quarried near Prestwick, Symington, Dreghorn, Kilwinning, Dalry, Kilbirnie, and Beith. Half-way between West Kilbride and Fairlie is Kaim's Hill, from which famous millstones are quarried.

Brick-clay and fire-clay occur in extensive beds on the surface near the coal measures around Hurlford, Kilmarnock, and Kilwinning, and these are largely used in making brick, tile, and sanitary ware. Fine potter's clay is got at Bentstone near New Cumnock.

Of the rarer minerals sulphate of baryta and porphyry occur at Misty Law. Lead and antimony have been worked in the parish of New Cumnock. In Galston chalcedony, agate, and other precious stones, known as the Burmawn or Galston pebbles, are found on the surface. Some years ago a thin vein of platinum was discovered on the rocky coast near Dunure. Copper ore of superior quality has been found on the estate of Ardmillan south of Girvan, where it is supposed to exist in considerable quantities ; it has also been wrought to some extent at Daleagles in the parish of New Cumnock. A few specimens of agates, porphyries, and calcareous petrifactions have been found among the hills of Carrick, and in the parish of Stair antimony and molybdena occur. The Water of Ayr whetstone is worked at Dalmore and at Bridge of Stair on the Ayr.

15. Fisheries and Fishing Stations.

The great increase in the capture of fish within recent times is due in no small degree to the facilities for transit now enjoyed, by which a supply of cheap and palatable food is readily available in the most remote parts of the country.

Fisheries, both sea and inland, have always been considered an important source of wealth to the nation, so much so that governments have taken measures to protect them by laws relating to ownership rights, as well as to the mode and time of capture.

The total weight of fish landed in England (exclusive of salmon and shell-fish), of which upwards of 80 per cent. are landed on the east coast, is over 520,000 tons, in Scotland above 320,000 tons, in Ireland 35,000 tons.

Salmon fishing is also important. Its value in Scotland varies from £200,000 to £300,000 annually. In Ireland the value is from £300,000 to £400,000.

The number of men and boys actively engaged in fishing from the United Kingdom amounts to 107,000. In England there is one fisherman in 612 of the whole population, in Scotland one in 116, and in Ireland one in 216. The total number of registered boats is close on 27,000 (1750 being steamers), of which about 11,000 are Scottish vessels.

The fisheries round the coast of Scotland, especially the east coast, are very important, and give employment to an army of people. No fewer than 35,000 persons are engaged in catching and curing the fish, and quite a multitude of people are employed in distributing over 3¼ million pounds' worth that reach the Scottish shores every year. These statistics take no account of the number of persons employed and the value of the fish obtained from inland fisheries in lake or river. Considerable outlays have to be made in vessels, machinery, and gear, all of which are subjected to much wear and tear, and are often exposed to the risks of total loss.

Fish that live and feed near the bottom of the sea as plaice, sole, cod, haddock are caught by the trawl-net, or by baited hooks on lines. The herring, which goes in shoals near the surface, is caught by drift-nets. Crabs and lobsters are taken in traps, usually dome-shaped cages of wickerwork or nets stretched over a strong frame. Other crustacea of some value as crayfish, shrimps, and prawns are caught in peculiarly shaped nets; and among the more important molluscs the oyster is taken by the dredge, while the mussel, the clam or scallop, and the

Ballantrae Harbour

whelk are found clinging to any rocks to which they can anchor.

Fishing is practised in some measure all along the extensive seaboard of Ayrshire, but Ballantrae, with Girvan, is by far the most important head-quarters. Girvan was long noted for its salmon fishing, but this industry has declined, and *white fishing* is now its chief industry, that is, the fishing for cod, whiting, soles, and flounders. South of Girvan there are one or two important lobster fisheries.

It would serve little purpose to name all the varieties of fish that are caught along the Ayrshire coast, but 30 different kinds have been reckoned up that are commonly, or occasionally, used as food. The most valuable are the turbot, the sole, and the halibut. The most abundant are whiting, haddock, cod, ling, flounder, mackerel, skate, and herring. Those less frequent are mullet, pilchard, lythe, and saithe. There is great abundance of conger eel, occasionally found in the trawl-net, and lobsters, crabs, and shrimps may be mentioned among the shell-fish taken.

In the lochs and streams there is an abundance of trout, and salmon is found in the larger rivers as well as in the Clyde estuary and in Loch Doon. For this variety of fish the Doon is the most productive of all the rivers. The char is found in Loch Doon and the miller's thumb in Carmel Water. The thick-lipped grey mullet is gregarious in some of the rivers. The perch, braize, and pike are common in lochs and sluggish streams, and in the shallows of these are the minnow and the beardie.

16. Shipping and Trade—The Chief Ports, Sea Trade Routes.

There is a considerable amount of traffic both local and over-seas along the Ayrshire coast. The chief sea-ports are Ardrossan, Irvine, Troon, and Ayr. Girvan has a small coastal harbour much improved since 1881, and twice extended within the last 50 years. Its accommodation has consequently much increased. Although it has some shipments of coal, limestone, and grain, its importance is trifling when compared with that of the other four places named.

The most northern port is Ardrossan. It possesses an important shipping connection which has greatly developed

The Harbour, Girvan

since the excellent harbour was enlarged and improved at great expense. The harbour owed its rise originally to Hugh, twelfth Earl of Eglinton, who conceived the idea of connecting Ardrossan by canal with Glasgow and Paisley and so making it the great commercial emporium of the West of Scotland. The canal got as far as Johnston from Glasgow, and the gigantic scheme of harbour construction, which was both tedious and laborious, was prosecuted at an immense outlay. The docks and quays occupy 17 acres. The harbour is sheltered by a natural breakwater called Horse Island. It is, therefore, not only very accessible, it is also safe. The port is of some consequence for passengers as well as for merchandise. A regular

At the Harbour, Ardrossan

service is maintained throughout the year to Belfast and Manchester, and during the summer months there is a steamship connection with the Isle of Man, along with a daylight service to Belfast and Portrush. Its chief imports are iron-ore, limestone, grain, and timber, while it exports large shipments of coal nd pig-iron.

The port of Irvine is under the authority of a harbour trust. In 1645 Irvine joined a small committee to consider the question of the India trade, and before 1760 it was a customs port ranking third among the ports of Scotland with a list of 77 registered vessels. Before the construction of Port Glasgow harbour it was largely used by Glasgow merchants for their sea-borne trade. The harbour, which has been much improved at very considerable expense, is safe for vessels from any wind. It extends to almost nine acres. The imports are mainly limestone, ores, salt, grain, timber; the exports are coal, pig-iron, fire-clay goods, and chemical products.

Troon possesses excellent docks constructed by the fourth Duke of Portland, and acquired in 1901 by the Glasgow and South Western Railway Company. Till 1860 Glasgow had to send her biggest ships to Troon for repair, as its principal graving dock was then the largest on the Clyde. Troon harbour carries on a considerable amount of foreign commerce, and its general coasting trade and its traffic with Ireland are of a fair extent. Its imports consist of wood, limestone, iron and other ores, and its exports are coal and pig-iron. A shipbuilding yard leased by the Ailsa Shipbuilding Company employs at times as many as 600 workers.

In early times the harbour of Ayr was the principal port on the Clyde estuary, and, in spite of some natural difficulties, such as a bar at the mouth of the river and a liability to be silted up by heavy floodings, it still enjoys a large share of the commerce of the county. By means of a powerful dredging machine kept constantly at work,

Ayr Harbour

the harbour is maintained in an efficient condition. It is a commodious one, formed by two long piers on each side of the river mouth, and protected by a breakwater. The trade is chiefly with Ireland, but considerable foreign and coasting traffic is also carried on, the exports consisting of cottons, woollens, iron, coal, whetstones, and paint; and the imports of grain, cattle, spirits, timber, slates, bricks, and lime.

17. History of the County—Main Features.

The early history of Ayrshire is dark and legendary. Of the intercourse, peaceful or warlike, of the early Celtic inhabitants with the Romans nothing authentic is known; and hardly anything trustworthy of their struggles with their neighbours—the Angles, the Scots, and the Picts—after the Romans left Britain. Ayrshire, however, was certainly part of the Welsh or British kingdom of Strathclyde, which had as its capital Alcluyd, afterwards called Dumbarton. In the eleventh century Strathclyde was merged in the kingdom of the Scots.

After the accession of Edgar, son of Malcolm, in 1097, strangers from the south came crowding into the country bent on its colonization. Grants of land were given them, and barons sprang up, who built castles and endowed churches, following the practice of England. Towns too arose inhabited by persons engaged in trade. These new communal conditions led to different customs and rights, and, consequently, about that time some changes were necessitated in the laws of the country for the administration of justice.

During the reign of Alexander III the Norsemen made repeated incursions on the western isles and coasts of Scotland, over which they claimed sovereignty. For long this was a serious evil. Not only was the presence of the vikings so near the mainland a constant source of danger in itself, but they leagued themselves with England and with the feudal lords of Galloway and Argyll when these were at war with the central power. The active Scottish king had driven out from the western isles and the islands in the Firth of Clyde all the Norse chiefs who refused to give allegiance to the crown, and in the autumn of 1263 King Haco of Norway fitted out a fleet to enforce his claims to the supremacy of these parts. Sailing along the Western Isles, the Norwegian fleet at length cast anchor between Arran and the Ayrshire coast. At Camphill, between Kilbirnie and Largs, on the heights overlooking the Ayrshire coast, lay the host of the King of Scots. A fierce storm diminished and broke up the Norse galleys. In the battle that ensued the Norwegians were driven to their ships, and Haco sailed away to die at Kirkwall. Alexander followed up his advantage; and three years later the Western Isles were united to the crown of Scotland.

Much of the story of Wallace is intimately associated with Ayrshire. His grandfather was Adam Wallace of Riccarton, and his father was a knight and landowner of Auchinbothie in Ayrshire, as well as of Elderslie in

Renfrewshire. Many places in the county are linked with incidents in his career, as Ayr, Irvine, the Laglain Wood, Ardrossan, and the region of Afton Water.

The son of the first Bruce became Earl of Carrick in right of his wife and succeeded to the ownership of Turnberry, which was afterwards regarded as a stronghold of the Bruce family. It was probably here that their son Robert Bruce was born, who on the expulsion of Baliol laid claim to the Scottish throne; and it was here that he struck the first effective blow for his rights. Ayr, the Wilds of Carrick, and Loudoun Hill have also much to tell of his wanderings and exploits.

Ayrshire may well claim to be a royal county. Carrick gave us the Bruces. Kyle cradled the Stewarts. Cunningham furnished the mother of the Stewart monarchs. Turnberry, Dundonald, and Rowallan form a trinity of royal Ayrshire homes. The castle of Dundonald was the favourite residence of Walter the first Lord High Steward of Scotland. The sixth Steward wedded Marjory, eldest daughter of the Bruce, whose only son Robert Bruce Stewart, Robert II, ascended the throne as the first of the Stewart kings. The first wife of Robert II was Elizabeth, daughter of Sir Adam Mure of Rowallan Castle.

Scarcely any district in Scotland remained so long under the baronial system as Ayrshire, and for centuries the chiefs were perpetually engaged in feudal broils and lawlessness, due in no small degree to the inefficiency of the crown to suppress them. Seldom did redress or punishment follow outrage, and owing to the weakness of the government the rival feudalists were generally allowed to fight out their own quarrels among themselves, which generally arose out of family jealousies and contendings for heritable jurisdictions, like stewardships and bailiwicks of districts. The famous feuds between the Montgomeries (Lords of Eglinton), the Cunninghams (Earls of Glencairn), the Boyds (Lords of Kilmarnock),

Rowallan Castle

and the Stewarts of Darnley in Renfrewshire, with each of whom many others joined, were long-standing and sanguinary. The southern part of the county was long distracted by hostilities between one branch of the Kennedys and another. In 1561 and subsequently, this powerful clan was involved in a fray of no ordinary importance arising out of the violent and fraudulent acquisition of church property by Gilbert, Earl of Cassillis, called in his own neighbourhood the "King of Carrick." Along with his brother, Thomas Kennedy of Culzean, he cruelly treated in the vaults of Dunure Castle the Commendator of the Abbey of Crossraguel until he got certain writs signed in his favour for the conveyance of the lands and property of this ecclesiastical foundation.

In 1678 a committee of the Privy Council sat in Ayr for directing the military executive in their effort to crush the covenanting movement. At the instance of the Council a horde of half-civilized, undisciplined, and savage Highlanders, possessed of all the predatory habits of their race, marched into Ayrshire and quartered themselves upon the people. They robbed upon the high road, plundered houses, and inflicted such punishments as they thought fit. But after a long and weary struggle the spirit of the people was firm; for, supported by the active sympathy of the nobility and the gentry of the county, they stood strong for their right, and the invasion of the Highland host failed in accomplishing its purpose. After the departure of the Highlanders the weary struggle dragged on. At one time fortune gave victory to the covenanting army at Drumclog, at another woeful disaster overtook it at Aird's Moss, where a number of worshipping peasants were surrounded by a troop of well-armed and well-disciplined dragoons, who had been scouring the Ayrshire uplands and western wilds. The moss-hags and caves and hidden recesses have many memories of these wild times, and the martyrs' graves that dot the moors and ennoble the village churchyards of the county are the sad memorials of the indomitable spirit that won. For long years after, the tradition of these harassing times continued, and influenced the people. In their minds the Stewarts were prominently identified with these long years of persecution. The inhabitants, therefore, not only gave ready accession to the government of William III, but in the Jacobite risings they also stood firm and active for established order.

The subsequent story of the county is one of progress. In all political movements for the amelioration of the people's lot it has taken its due share. In the agitation of 1832, and especially in the chartist movement of 1838, the men of Ayrshire occupied a position of outstanding prominence.

18. Antiquities—Prehistoric, Roman, Celtic.

In the absence of all written records much light is thrown on the activities of primeval savage man by remains of his handiwork which have reached us from early times—the arms, implements, ornaments, coins, camps, forts, standing and sculptured stones, dwellings, and burial-places of ancient races. From the study of these mute memorials we get little information as to exact spaces of time within which each is included. It is enough that broadly three successive stages, marking the degree of civilization attained, are recognized—the Stone, Bronze, and Iron Ages. But, though an article may be of stone it is not difficult to conceive that it may have continued in use long after the introduction of bronze, and it is well known that bronze was employed long after iron came into general use. This makes it more difficult to determine to which age each relic really belongs.

Remains of the Stone Age have been found in various parts of the county. A flint arrow-head, the edges finely

Ring of Jet, found in a Cairn at Mosside

serrated, was found in moss near Newmilns; another at Dalbeth. A drill or borer of flint was found in Galston parish; and in the same district, under seven and a half feet of peat, a whorl of felstone. Near the old castle of Trochrigg there was unearthed a large stone mould or matrix for casting metal objects of unknown use; and near Tarbolton a celt. A flint scraper and various fragments of worked flint, together with some articles in metal and pottery, were dug up near Maybole; and on the shore between Troon and Irvine some stone arrow-heads were discovered.

The relics of the Bronze Period show a much higher state of civilization than those of the Stone Period, but these are rare in Ayrshire. One of the chief specimens is a bronze shield, $23\frac{3}{4}$ inches in diameter, found about

Arrow-heads of Flint, found at Lanfine, Newmilns

the year 1779 in a peat moss on the farm of Lugtonridge near Beith. It is now in the possession of the Society of Antiquaries in London. Four or five other shields are said to have been found at the same time, but this specimen appears to have been the only one preserved. These round shields of bronze are said to be the most remarkable of personal relics of warfare in Scotland. They are beautiful and accurate in construction, and from the device of a number of concentric circles their outer surface has the appearance of being richly decorated. From the nature of the material used in their construction they probably date back to a period before the time of the Romans. At Monk, near Galston, there was also discovered a tripod pot of brass ornamented on the upper face with incised concentric circles. Trinkets and other relics of bronze, popularly supposed to be evidences of Roman occupancy, have been found in different parts of

Arrow-head of Flint found at Lanfine, Newmilns

Borer of Flint found at Changue, Galston

the county—various vessels, a tripod, a spear head, and a small image representing Justice with equal weights. Care must be taken, however, not to conclude that the discovery of these relics always shows that Romans were their last possessors. Being small they might easily have been picked up by travellers, and, after being shifted from one place to another, they might have conceivably been dropped on the way.

If we except the old road from Ayr to Dalmellington and Loch Doon, which is claimed as Roman, undoubted relics of Roman camps and buildings are non-existent in Ayrshire.

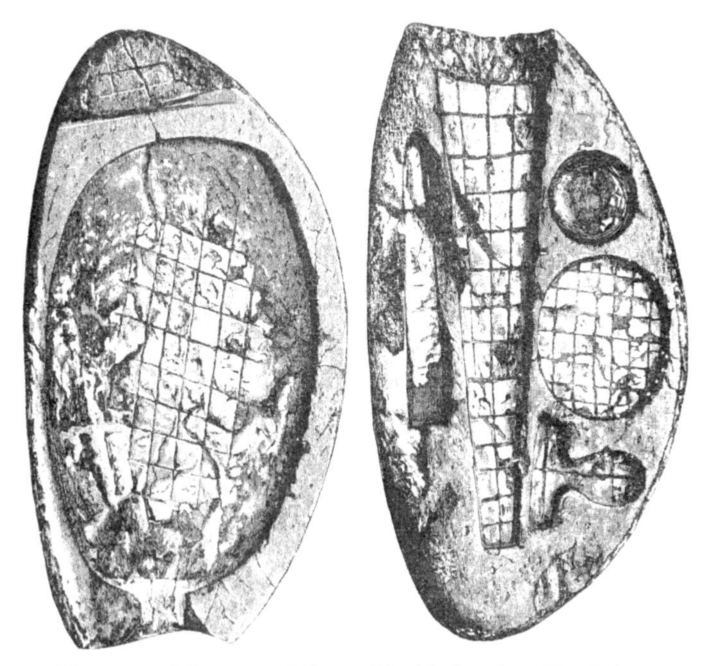

Obverse and Reverse of Stone Mould, found at Trochrigg

Of other ancient structures Ayrshire has numerous specimens in the form of motes, mounts, castle hills, and curvilinear forts.

The motes generally conform to the true type, having

an elevated mound-like character, with circular or oval form and flattened top, sometimes with a rampart or

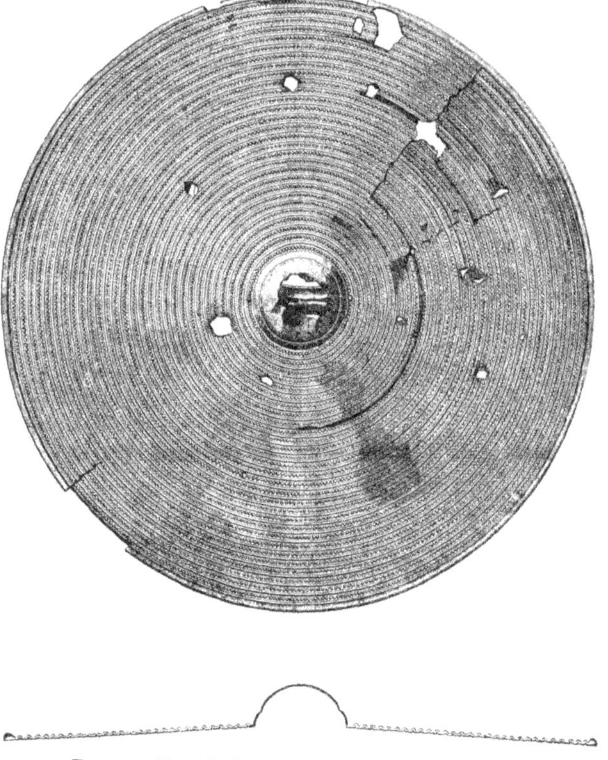

Bronze Shield, found at Lugtonridge, Beith

trench thrown up for defence. The most important of these are at Dalmellington, Kilkerran, Alloway, Monkwood, Girvan, Old Cumnock, New Cumnock, Colmonell,

Bennane Head, and Straiton. It is noteworthy that all these are in the southern half of the county. In the northern half there are only a few apparently artificial "mounts" with the usual flat oval summits, but with little trace of fortification. The chief are the Glen Mount, West Kilbride, and Knock Rivock Mount, Ardrossan.

Tripod Brass Pot, found at Monk, Galston

These should perhaps be more properly classed with the moot-hills, law-hills, or court-hills where the barons or their deputies met to dispense justice, and whither the people repaired to confirm contracts and to determine controversies publicly in face of the sun. They are found in various parts of the county.

Of the structures certainly or doubtfully identified as forts 31 can be proved to exist or to have existed although six of them are now without the mark of a site. Unlike the motes their distribution is not confined to the southern half of the county, but they are limited to a strip running from Camphill in the north to Finnart Bay in the south, and all are only a short distance from the sea. They are of the usual circular or oval form peculiar to the British or Scandinavian type, with from one to three lines of defence. The fort of Caerwinning, which is

Dalmellington Mote

probably the largest, lies two miles from Dalry, on a conspicuous height commanding an extensive view. It appears to have been a place of great strength and accessible only on the west, where the entrance seems to have been. Parts of the three enclosures of stone usual to prehistoric British camps are distinctly traceable. In the Dundonald district there are three forts—Harperscroft, Kemplaw, and Wardlaw.

Of other ancient relics none throw a fuller flood of light on the little-known manners and customs of prehistoric life than the crannogs or lake-dwellings. These are constructed on artificial or partly artificial islands in lakes used by the early Celtic inhabitants as places of refuge and defence. They are constructed with much ingenuity, and are found in places without natural protection. These crannogs are interesting not only from a consideration of their structure but from the various relics found in them, showing that the inhabitants were artificers of some skill. The most remarkable of those discovered in Ayrshire was that at Lochlee in the parish of Tarbolton. It was systematically explored, and a great number of relics were discovered indicating that the people had made considerable advances in civilization. The discoveries made in the excavation of a crannog at Lochspouts in Maybole parish, of one at Buston near Kilmaurs, and lastly of one at the south-west corner of Kilbirnie Loch were also of considerable archaeological importance. At Buston the remains of a dwelling-house were distinctly discernible. That the inhabitants were skilful carpenters and that they possessed, if they did not make, articles of bronze and other metals is proved from the discovery at Loch Doon of several ancient canoes, in one of which was a battle axe; and of another at Kilbirnie Loch containing a bronze tripod pot and a lion-shaped ewer, objects probably belonging to some century between the sixth and the tenth.

The story of the respect which the living invariably showed to the dead may be read in the tumuli discovered here and there containing urns with calcined human bones. Excavations of these have been made at a place called the Moat, near Ochiltree, and at Seamill, near West Kilbride. Stewarton parish has been rich in these unknown graves. Kilbirnie, too, had an interesting tumulus, and a good many urns and other ancient relics were turned up on the Sandy Knowes of Wallacetown. Stair, Stewarton, Beith, Dalry, and other parishes have also contributed their quota to these chambered relics.

In the parish of Kirkoswald a stone coffin was unearthed and found to contain some curious ornaments; near Ballantrae across the river from the ruins of Ardstinchar Castle there stand the so-called Druidical stones of Garleffin. On the northern declivity of the Cuff Hill, Beith, there is a stone of considerable size, said at one time to have been able to be set in motion by the slightest touch. In the space between other four stones several remains of great antiquity have been exhumed, including calcined bones and earth, which were supposed to mark the spot as a place of sacrifice. At the bottom of the hill north of the "rocking-stone," at a place called *Kirklie* Green, were remains of a trench and some buildings. Under a large cairn at the south bottom of the hill several stone coffins were discovered, fragments of human bones being found in one of them, but the coffins did not lie east and west after the manner of Christian burial. Again, on the adjoining lands of Threepwood a large vase of burnt clay was unearthed having in it a quantity of calcined bones. The structure of the vase, which seemed to be formed by the finger and thumb without any aid from the potter's wheel, indicated no knowledge of art. The stone and the cinereal urn show that the two forms of burial are sometimes met with together, and they probably indicate two different ages remote from each other.

Specimens of sculptured stones are not common in

The Hunterston Brooch

Ayrshire. The only examples of these monuments of the early Christian period were some stones discovered at Mansefield House, near New Cumnock railway station.

These were sculptured in relief with Celtic ornamentation but without symbols.

A most interesting relic of late Celtic art is the Hunterston brooch, discovered on the Hunterston estate, and now in the Antiquarian Museum at Edinburgh. It is of silver richly ornamented with gold filigree work and set with amber. In artistic excellence it is the finest of the penannular brooches found in Scotland, and almost equals the famous Tara brooch. The Hunterston brooch has runes scratched on the back, indicating two owners—a man with a Celtic name, a woman with a Scandinavian. Brooch and runes belong to the Western Isles about the tenth century.

19. Architecture—(*a*) Ecclesiastical.

The architecture of the buildings in Ayrshire will be considered under four divisions, viz. (*a*) Ecclesiastical, or buildings relating to the church; (*b*) Military, or castles and peels; (*c*) Municipal, town-halls and other public buildings; (*d*) Domestic, famous seats.

Let us consider first the churches, and then glance at the remains of abbeys, monasteries, and other religious houses. The churches of Ayrshire are of various styles and of different ages, so that it may be well to classify them here as Old Scottish, Early English or English Gothic, and Perpendicular. Examples of all these styles or developments of them are to be seen in Ayrshire.

It is not uncommon to find specimens, especially in the smaller towns, of the Old Scottish type of architecture with their corbie-stepped gables, steep roof, quaint windows, and small square tower crowned with a small belfry. Aisles may or may not be attached. These are sometimes added as demands for accommodation arise. One of the best examples of this style is the ancient parish church of Kilbirnie, part of whose construction must date back to the end of the fifteenth century. This church is chiefly remarkable, however, for the wonderfully profuse oak-carvings with which the Crawford Gallery and the pulpit are decorated. It is possible here only to generalize, but the whole is a mass of ornamental carvings with pediments, columns, and entablatures. The details of this extraordinary display show heraldic devices characteristic of early Scottish work, enriched with scrolls, wreaths, foliage, and flowers. The old "parochial church" of Beith, of which there now remains little more than the south or front gable and belfry, was of the same type of architecture. A stone in this gable bearing date 1593 would seem to have belonged to an older building, which was taken down when the church was rebuilt and enlarged. These are only specimens of a style of building found in various parts of Ayrshire.

The style called Gothic or Early English exhibits much grandeur and splendour, and is characterised by its lofty pointed vaults, its pinnacles and spires, its large buttresses, its steep vaulted roofs, its long narrow lancet-headed windows, and, generally, by its lofty bold character. This type of building arose from the endeavour to cover the widest and loftiest areas with the greatest economy of stone. It has got several varieties representing different

Front Gable of Old Church, Beith

epochs; and there are few towns without examples of modern churches of this build, whose spires are not only objects of approbation, but also pleasing landmarks. These are frequently seen more particularly in the larger towns of Ayr, Kilmarnock, and Irvine, as well as in Troon, Girvan, Saltcoats, and Largs. Perhaps one of the finest specimens of a modern edifice in the style of thirteenth century Gothic is the Episcopal Church of Holy Trinity at Ayr, which consists of nave, with north and south aisles, chancel, organ loft, clergy and choir vestries.

The Perpendicular style without much ornament is very prominent in the county, more especially in the case of the parish churches, which are of a familiar type. It takes its name from the general vertical tendency of its tracery. Perpendicular lines prevail in the windows, which are furnished with transoms or cross-bars as well as mullions. Gables and roofs are generally at a low angle. A good example of this type is Beith Parish Church, with its lofty square tower divided into five unequal portions by moulded string-courses, and terminated by four pinnacles at the corners with a characteristic parapet intervening.

The existence of powerful monastic institutions in the county shows that as early as the fourteenth century the ecclesiastical life of Ayrshire was very highly organised. The chief centre was at Kilwinning, where Hugh de Morville, of Norman descent, erected about 1140 a splendid abbey, whose ruins still stand, dedicating it to St Winnan, an Irishman of the eighth century. It is said to have been "solid and great, all of freestone cut; the church fair and stately after the model of that of

Glasgow, with a fair steeple of seven score foot of height." What remains of the gable of the south transept in the lancet style, which is still nearly complete and calculated to have been 80 or 90 feet high, together with an arch leading into the nave, shows it to have formed part of one of the noblest examples of Gothic architecture raised in Scotland.

The Abbey, Kilwinning

The second of the great monastic institutions in the county was the abbey of Crossraguel. The side walls of church and choir still remain to the height of 14 feet. The abbey was partly an old baronial keep, partly an ecclesiastical fabric; and a lofty defensive tower, which was the usual bulwark of an ancient peel, overlooked the pointed pinnacles of Gothic architecture. After the manner of Glasgow Cathedral it was arched with vaults and cells. Duncan, Earl of Carrick, to whom the whole of the southern division of the county had been apportioned, founded the abbey in 1244, endowed it with some of his lands, and granted to it the patronage of a number of churches in the vicinity. Its endowments were supple-

Crossraguel Abbey, Choir and Chapter House

mented by grants from Nigel, Duncan's son, and the Bruces of Turnberry: and King Robert erected it into a free barony with full jurisdiction over the people. Nearly the whole of the county south of the Doon owed allegiance to the monks, who consequently maintained a strong hold on all public affairs, enforcing their sovereign temporal power by spiritual authority.

At Mauchline there was a religious establishment, most likely a priory, being an offshoot from the monastery of Melrose. The only part of the building remaining is the tower called Mauchline Castle, an old square edifice with crow-stepped gables and breastwork fortification.

A settlement of the Trinitarian order of Friars for the redemption of Christian captives, sometimes called Red Friars, existed at Fail. Of the original monastery there still stand the gable and part of the side wall of a house. The inmates are satirised in the lines—

> "The Friars of Fail
> Gat never owre hard eggs, or owre thin kail
> For they made their eggs thin wi' butter
> And their kail thick wi' bread:
> And the Friars of Fail, they made gude kail
> On Fridays when they fasted.
> And they never wanted gear enough,
> So long as their neighbours' lasted."

The Dominican or Black Friars first established themselves in Scotland at Ayr in 1230. The monastery stood on the bank of the river Ayr near that of the Grey Friars founded in 1472.

The Convent of White Friars or Carmelites in Irvine was founded by Fullarton of Fullarton in 1240, and continued to flourish till the Reformation. The last prior, seeing the threatening storm, alienated the lands, which went under the name of Friars' Croft. In a charter granted by James VI in 1572 the revenues of the order were applied to the erection of a school, from which through time arose the town's Royal Academy.

In Maybole there are still preserved the ruins of a collegiate church founded by Sir John Kennedy in 1371.

There are three roofless ruins—all pre-Reformation churches—Alloway Kirk, Monkton Kirk, and Lady Kirk. Alloway's "auld haunted Kirk" is supposed to have been erected about the year 1516. Burns made it

Auld Alloway Kirk, Ayr

the scene of the witch-dance in *Tam o' Shanter*. Some hold that the Saxon arch in the church of Monkton indicates a time long prior to the Reformation. In fact, Blind Harry associates it with the scene of Sir William Wallace's remarkable dream. Lady Kirk stands at the north-east corner of the parish of Monkton and Prestwick.

According to Chalmers in his *Caledonia*, "The building formed a square, having turrets upon each corner; and there was a chapel in the middle of the square. The chapel was dedicated to the Virgin Mary, from which it obtained the popular name of 'Our Lady Kirk.'...In a grant of James IV, in 1505, it is called 'the preceptory of our Lady Kirk of Kyle.' There appears to have been connected with this establishment a *Pardoner* who was popularly called 'Our Lady of Kyle's Pardoner.'"

When we look at the place-names of Ayrshire, more especially in the western half, we are struck with the number of *Kils*, *Kirks*, *Crosses*, *Ladys*, etc., all which testify to ecclesiastical foundations of various ages. *Kil*birnie, *Kil*winning, *Kil*bride, *Kil*maurs, *Kil*marnock, *Kil*kerran; *Kirk*oswald, *Kirk*michael, *Kirk* Dominae (near Barr), *Kirk*cudbright-Innertig (the ancient name of the parish of Ballantrae); *Cross*hill (4), *Corse*hill, *Cross*holm, *Cross*craig, *Cross*house, Portin*cross*, *Corse*, *Cros*bie (3), *Corse*house, Lady *Corse*, *Cross*raguel are names which are merely the impress of that seal. In Monkton parish was Our *Lady* Kirk of Kyle, Prestwick has its *Lady*ton, and off Troon harbour is *Lady* Isle. On the lands of Crosshill in Kilwinning parish is *Lady* Acre. In Tarbolton there is a *Lady*-yard, in Kilbirnie a *Lady*-land. Connected with the "Auld College" of Maybole were a *Lady*land, a *Lady*well, and a *Lady*corse. Ayr, too, had its *Lady*landis, Kirkoswald an estate called *Lady*-bank, and Dailly its *Lady*glen. In Dunlop the *Lady*steps were the steps that crossed the burn to the Chapel House.

20. Architecture—(*b*) Military. Castles.

There is nothing to show that Scotland was subject to the same influences as produced the grand specimens of Norman castles whose stately ruins still exist in England. The inhabitants had to think how they were to defend themselves not only against a common foe, but more commonly against the attacks which one feudal family was constantly inflicting on another. They consequently built for themselves peels, keeps, and lofty towers in places least susceptible to attack and most easily defended.

These Scots peel towers were generally raised on the brink of a perpendicular rock rising high out of the sea or river, or on the summit of a round hill difficult of access, with or without a fosse or ditch on the land side. Others took advantage of such natural protection as an island on a lake offers. Others, again, were built on comparatively level ground with the security furnished by water collected by damming up a stream, or by some other artificial means of defence. Wherever they were situated they were all dark and cheerless. The walls were thick, the doors and windows small. Serving the purposes both of a home and a fortress, they point to an age of tumult and rapine, when comfort according to modern ideas was only a secondary consideration. The living quarters were mere cells set in a solid mass of masonry. The only specimen in Ayrshire of a castle making some attempts of a decorative character and some provision for comfort, was that at Maybole, the ancient capital of Carrick, where the Lord

Cassillis and other local magnates held on occasion a miniature court.

"Carrick is the kingdom of castles. Along the sea-shore, up the valleys of the Doon, Girvan, and Stinchar, everywhere they are to be found ruined or renovated, each with a story of romantic interest." Of the ruined

Maybole Old Castle

castles by the Doon that which stands on an island near the head of the loch is one of a group of ancient keeps round which much of the history of the shire is entwined. Its ruins and position are an evidence of strength. Most of the Carrick families are connected with the castles in the Girvan valley—the Kennedys with Bargany, Blairquhan, and Dalquharran; the Mures with Cloncaird, now the home of the head of the Wallaces; the Cathcarts with Killochan; the Boyds with Penkill; and the Fergussons with Kilkerran. The towers on the Stinchar are Ardstinchar, Knockdolian, and Craigneil. Ardstinchar from its position commanded not only the ford of the river but the two entrances into Carrick, that along the shore, and that which leads up the river and across the country. At the base of the cone-like hill of Knockdolian is the hoary old ruin of the same name, ivy-clad to the turrets. Across the Stinchar from Colmonell is the castle of Craigneil, a grim old peel, rearing its bare walls above a rocky eminence.

On the shore there is no castle of greater historic interest than that of Turnberry. All that is left of it is a few broken walls standing on a rocky point where the coast from the south curves inward towards Girvan. The memorable part it played in the troublous time when Bruce was its lord, is evidence that it was a place of great strength. Dunure Castle, also on the coast further north, stands on the very verge of the sea. It has been a ruin for over 200 years, and for 400 years was one of the strong keeps of Carrick, the principal seat of the ancestors of the house of Kennedy. It was the scene of the gruesome story associated with the torture by the Earl of Cassillis of Alan Stewart, Commendator of Crossraguel Abbey. Other castles in the same district are Greenan and Dunduff.

In Kyle the castles are not so numerous, but some of them are of great historical interest, others were of considerable strategic value. Dundonald Castle, on an isolated picturesque knoll, is a grim ruin in the very old Scottish order of architecture. It is a square tower of several storeys, each of which consists of only one room, the walls being of prodigious thickness. The leading feature of this castle is the great arched hall, 58 feet by 24 feet, with the arch 40 feet above the present floor. Dundonald was the home of Robert II; and "Dr Johnson," says Boswell, in his *Tour*, "to irritate my old Scottish enthusiasm, was very jocular on the homely accommodation of 'King Bob,' and roared and laughed till the ruins echoed." A little north of Ochiltree on an almost inaccessible height is the old castle of Auchinleck, the home of the Boswells, with whose sullen dignity Dr Johnson was greatly delighted. Other ancient castles of the middle region of Ayrshire were Kyle, Craigie, Ochiltree, Auchincloigh, Laight, and Riccarton.

Cunningham has no fewer than 17 old castles, some of them being still inhabited, so that it might well dispute the claim already made for Carrick. These are Seagate (Irvine), Dean, Craufurdland, Rowallan, Giffen, Hessilhead, Glengarnock, Kilbirnie "Place," Monkcastle, Kerilaw, Ardrossan, Law, Crosbie, Portincross, Hunterston, Fairlie, and Knock. Some of these, like Craufurdland, Crosbie, and Monkcastle, having been partially rebuilt, occupy sites of ancient edifices and are at present inhabited. Dean Castle was long the residence of the Boyds, Earls of Kilmarnock. Its roofless ruin has just been entirely renovated by the proprietor, Lord Howard de Walden. Rowallan, a preserved ancient ruin and perhaps as picturesque as any of the numerous baronial castles of Ayrshire, three miles north of Kilmarnock, is a perfect specimen of an early feudal residence. Of the original fortlet only the vaulted under apartment remains. This tower has been assigned as the birth-place of Elizabeth Mure, the first wife of Robert (of Dundonald) the High Steward,

Hessilhead Castle from the North

afterwards King Robert II. Hazelhead (Hessilhead) Castle, a strong old building seated on a loch and environed with ditches (all long since drained and levelled), was the seat of a branch of the Montgomery family. It is about two miles east of Beith, and is believed to have been the birth-place of Alexander Montgomery, early Scottish poet. It is now an ivy-clad ruin surrounded by some fine old trees planted towards the end of the seventeenth century. Kilbirnie Castle, the "Place" or "Palace" as it is locally called, is a ruin a little to the south-west of the town of Kilbirnie. About 1395 its earliest known residents were Barclays, a branch of the Barclays of Ardrossan Castle. Near Farland Head, on a ledge of rock within the sea

"The Place," Kilbirnie

wash, are the romantic ruins of Portincross Castle, an ancient and interesting relic with walls still pretty entire. It was a frequent residence of the early Stuart kings, who used it as a halting-place when travelling between their castles of Dundonald and Rothesay.

21. Architecture—(*c*) Municipal.

In Ayr the most commanding object is the steeple of the town buildings pointing upward, sharp and high, to the height of 226 feet. At the time of its erection it was considered the highest spire in Scotland. Other important edifices are the domed county buildings, with imposing frontage in Wellington Square, modelled after the temple of Isis in Rome, and the so-called Wallace Tower, a Gothic edifice 115 feet high.

Kilmarnock has its town house built on a bridge over the Kilmarnock Water near the centre of the town, and a Corn Exchange with its Albert tower 110 feet high.

The town house of Irvine is a large and handsome edifice with clock tower and belfry, terminating in an octagonal steeple, surmounted by a vane which rises in its front centre. The crown, sword, and sceptre, which were carved in relief above the fireplace in the court-house of the Old Tolbooth, find a place in the wall of the vestibule, with a lion rampant over them.

There are also beautiful town halls at Old Cumnock, Ardrossan, and Dalry.

In front of the town house of Prestwick is the "Mercat Cross" of the town, one of the best preserved specimens in the country. Mention is made of it in the burgh records as early as 1473, and in all likelihood it was in existence long before that year.

The Old Mercat Cross, Prestwick

22. Architecture—(*d*) Domestic. Famous Seats.

Of the specimens of domestic architecture in Ayrshire there are few that can date as far back as the War of Independence. Most of the earliest are of the sixteenth century.

The early domestic houses in Scotland were for the most part built of wood or wattles. The latter process was to have a stout double framework of wood interlaced with twigs after the manner of basket-making. The space between, which could be made as wide as might be necessary for the desired thickness of walls, was then filled with turf or clay.

The next stage was to build houses outside towns of turf or clay according to the material which the district most readily produced. These were covered with a roofing of tile or thatch. Ayrshire has now few specimens of the "auld clay biggin," but houses of the same material in a modified form may be seen in the long monotonous rows of miners' houses. Clay in the form of bricks is used in the construction of these because the material is generally accessible. Structures of this kind can be put together quickly, and their removal can be easily effected whenever the occasion demands. The chief material now employed in building the houses both in town and country is sandstone, carboniferous limestone, or whinstone, according to the geological character of the district, the purpose of the structure, the facilities for transport, or the wealth and tastes of the owner.

Penkill Castle

The houses of the nobles and gentry in the county afford as fine specimens of domestic architecture as may be seen anywhere; and when it is considered that the skill and good taste of the architect are often supplemented in no small degree by the charm of mount, or stream, or wood, or sea it will be admitted that they are mostly

Blairquhan Castle

picturesque and attractive residences. When the fighting spirit of the fourteenth, fifteenth, and sixteenth centuries ended there was no longer need for those gloomy fortifications, whose grim skeletons profusely dot the face of the county. In their places there sprang up family residences, in the construction of which suitable accommodation and comfort were the main concern. In these Ayrshire is especially rich.

Some of the domestic buildings in the county are merely outgrowths of the old baronial fortalice, which by repeated reconstruction and frequent repair in the course of the years have become commodious and comfortable

Killochan Castle, from South

residences still in good keeping. Examples of the modernised keep are Newark Castle on the Doon, Craufurdland Castle and Dean Castle near Kilmarnock, Sorn Castle, the Castle of Skelmorlie, Cloncaird Castle on the Girvan, and the Castle of Penkill. The Castle of Penkill was originally a high, square tower or peel, perched on the side of Penquhapple Glen, three miles east of Girvan. From the ornamentation in one of the windows, the original part is thought to date back to the fourteenth century. Considerable extensions were made in 1628 and it is now one of the oldest inhabited houses in the west of Scotland. Dante Gabriel Rossetti was a frequent visitor, and here he composed many of his poems.

Cessnock Castle

Of the domestic dwellings in the county which have not been reconstructed, the oldest known is the baronial mansion of Blair House, near Dalry. This structure is both unique and antique. It is in the form of two sides of a square, a style which suggests precaution against possible trouble. A rectangular projecting tower at the re-entering angle contains the staircase and main doorway The walls of the original structure, which has received various additions, are of great thickness. Few genealogies in Scotland can match that of the present laird, for the records of his house go back to William the Lion, and the family that he represents has dwelt at Blair in unbroken succession for 700 years.

Culzean Castle

Blairquhan Castle, a mile west of the village of Straiton, is a magnificent edifice most romantically situated. It occupies the site of the old keep of the Kennedys, and some of the lintels and sculptured stones of the old castle are built into the new. The ornate Tudor style of architecture has been most successfully carried out. Other ancient seats still inhabited are Killochan Castle on the Girvan, built in the baronial style of the sixteenth century, and the House of Cessnock near Galston, built in the French château style.

Other county residences, which can be little more than mentioned, are Culzean, Dumfries House, Eglinton

Eglinton Castle

Castle, Loudoun Castle, and Kelburne Castle. For situation few homes in the kingdom can surpass Culzean. Seated on the verge of a steep basaltic cliff 100 feet high overhanging the sea, it presents a range of lofty castellated masses that harmonize most admirably with its rocky site. The house itself is a modern imitation of the Gothic style. The main portion was built in 1777 on the site of an old keep called the Cove, so named from the caves or coves which extend under the rock on which the castle stands.

The Keep, Loudoun Castle

23. Communications—Past and Present. Roads, Railways.

Roads have been a gradual growth, developing from track-ways to by-ways, and from by-ways to high-ways. In the middle of the eighteenth century there was hardly a serviceable road in the country. Even then pack-horses were often preferred to wheeled vehicles as a means of transport. Stage coaches lumbered along where practicable at the rate of six or eight miles an hour. Towards the end of the same century civil and political contentions had given place to peaceful enterprise and social progress. Life became keener. Old easy-going methods had to yield to those that were newer and more alert. Improved roads were a prime necessity, and the two northern districts of the county were soon opened up by lines of communication in all directions.

McAdam, who was associated with Ayrshire by birth and for some time by residence, began his successful experiments in road-making in the county. All the roads are now constructed on the most approved principles, and are carried in the direction which necessity and convenience most demand. The character of the roads is largely determined by the material available for their construction. In Ayrshire they are macadamised with some variety of trap rock, usually that called whinstone. Hard limestone, where that is abundant, is likewise in use and forms a smooth and pleasant surface.

Three main roads, entering the county from the north, run from Glasgow, from which all the chief highways in the west radiate. The most westerly by Paisley passing through Beith, Dalry, and Kilwinning joins at Irvine the middle one, which runs through Barrhead and lies along the Lugton valley. The third reaches Kilmarnock by way of Fenwick. The two roads meeting at Irvine are joined by a shore road from Skelmorlie, Largs, Ardrossan, Saltcoats, and Stevenston; and their united course now keeps through Monkton and Prestwick to Ayr.

Three main roads also radiate from Kilmarnock—one east up the Irvine valley and thence into Lanarkshire, passing Hurlford, Galston, Newmilns, and Darvel; a second south-east past Hurlford, Mauchline, Auchinleck, Old Cumnock, New Cumnock, and on to Dumfriesshire; and a third west of south through Symington to Monkton, where it joins the Irvine and Ayr road. A short branch runs to Irvine by way of Crosshouse and Dreghorn, while another proceeds almost due north through Kilmaurs, Stewarton, and Dunlop and joins the middle artery to Glasgow at Lugton near the northern frontier of the county.

From Ayr a highway passes up the Ayr valley into Lanarkshire by way of Whittlets, Mauchline, Sorn, and Muirkirk. A second runs south-east past Hollybush, Patna, Waterside, and Dalmellington into Kirkcudbrightshire. A third proceeds by way of Maybole to Girvan where it meets a shore road, also from Ayr, winding round the bold and rocky coast past Dunure, Culzean, and Turnberry. From Girvan the main road continues along the shore to Ballantrae and on through Glenapp to Wigtown.

Besides these there are, of course, innumerable minor channels of communication which cannot find a place here. Probably one of the oldest highways in Ayrshire is the old road between West Kilbride and Dalry. It winds by a steep ascent up the western slope of Law Hill, and doubtless in days gone by it formed an important route for the traffic between the coast and inland parts.

The county is also well served by railways. A railway belonging to the Glasgow and South Western

Company enters the county in the north-east near Beith, runs to Dalry, whence a branch goes by way of Kilwinning to the coast at Ardrossan and on past West Kilbride and Fairlie to Largs. The main line continues from Dalry to Kilmarnock, Mauchline, and New Cumnock, leaving the county at a point about three miles east of the last named place for Dumfries. Branching off the main line at Kilmarnock, in a south-westerly direction a short line runs to Barassie and thence on to the extremity of the promontory on which Troon stands. It is interesting to note that this was the first railway constructed in Scotland, having been opened, with horse haulage, in 1812. Other branches run off in all directions to numerous towns and villages.

The main line between Glasgow and Ayr by way of Dalry, Kilwinning, and Irvine continues from Dalrymple past Maybole and Girvan on its way to Stranraer. An additional railway connecting Ayr with Girvan has been constructed along the shore past Dunure and Turnberry.

Two lines from Glasgow enter the county near Lugton, where there are two stations about 250 yards apart. From the more westerly of these stations there are two branches of what is called the Joint Lines, one west to Beith, and another south to Kilmarnock, and this section, managed jointly by the Caledonian and Glasgow and South Western Railway Companies, is used as the main line of the latter company from Glasgow to London. The other line at Lugton, entirely controlled by the Caledonian Railway Company, passes Giffen Junction and Kilwinning Junction and terminates at Ardrossan.

24. Administration and Divisions—Ancient and Modern.

Previous to the abolition of the feudal system Ayrshire was divided into the three districts of Cunningham, Kyle, and Carrick, each forming a separate jurisdiction of regality. Over these regalities authority was exercised by local potentates, called barons—representatives of the crown, who had the power to appoint deputies or baillies. The baron retained in his hand many instruments whereby he kept the vassals of his district enslaved, exercising in some of his feudal exactions a kind of petty despotism. The judicial president of the county was the "shire-reeve" or sheriff, the official deputy of the crown for the enforcement of law and order. This office was generally vested by hereditary right in some leading landowner unpossessed of any legal qualification. These powers were gradually being reduced until, by the Act of 1747, hereditary jurisdiction was abolished, and this proved the death-blow to the feudal law in Scotland. These bailiaries are now merged in one county administration.

Another jurisdiction was that of the Corporation, the highest kind of which was the Royal Burgh, a corporate body created by a charter from the crown. Of these Ayr, Kilmarnock, and Irvine are the only instances in the county. There was another kind of municipal corporation, the burgh of barony, of which Kilmaurs, Maybole, Newmilns, and Prestwick are examples. This was a tract of land created into a barony by the feudal superior and placed under the authority of magistrates elected either by the superior of the barony or by the inhabitants, according to the terms of the charter. Besides, there are several other towns known as police burghs, being populous places, with boundaries fixed in terms of statute, whose affairs are managed by commissioners.

Perhaps the oldest charter that absolutely brought a burgh into existence was that conferred on the town of Ayr in the thirteenth century when William the Lion was resident in the castle there. In 1250 Irvine also became a royal burgh. The franchise at first was much restricted and not very representative. Many years passed before it broadened down. But the right of voting in every variety of burgh is now extended to a much greater number of qualified citizens.

It was in 1535 that Ayrshire, in common with the rest of Scotland, first had the advantage of a Poor Law Act, which ordained that the infirm or incapable poor should be maintained by a tax levied on the parish. A similar provision obtains at the present day.

The various departments of public life in the county are now managed by specific authorities. The two chief officials are the Lord-lieutenant and the Sheriff. There are also a vice-lieutenant and thirty-eight deputy-lieutenants, but, apart from these, the general business is managed by an elected county council. This is a small parliament consisting of 57 members, with considerable powers exercised over the whole shire. It can levy rates and borrow money for public works. It keeps in repair the main roads and bridges, manages lunatic asylums and hospitals, controls police, appoints officers of health, and discharges other duties of county government. More immediate local control is exercised by the district councils, of which there are four in Ayrshire. These consist of the members of the county council for the division with one member from each parish council in the district. They are responsible for sanitation and water supply. The affairs of each parish, with respect to cleaning and lighting, are managed by a parish council. But its most important duty is to administer the Poor Law.

For purposes of education boards were instituted in all the parishes and in several burghs to adopt measures for providing efficient and suitable instruction for every child of statutory school age within their districts. These boards have power to borrow money for the building and the equipment of schools, and each has the right to demand from its parish council an annual sum to be levied by an imposed rate for the repayment of loans and the efficient maintenance of the schools. Besides, there is an Education Committee with the oversight of all secondary and technical education in the county.

For the administration of justice the county court is presided over by a sheriff. But as the sheriff is nearly always a practising advocate resident in Edinburgh, sheriff-substitutes act for him as local judges. Of these there are two in Ayrshire, one having his court at Kilmarnock the other at Ayr. There are also the Petty Sessions presided over by burgh magistrates or local justices of the peace to try offences in a summary way and to determine misdemeanours. The Quarter Sessions are held four times a year in the county town by the justices of the peace, with the power of reversing, if necessary, the sentences of the Petty Sessions when the finding is of a nature subject to review.

Ecclesiastical affairs are in the hands of Presbyteries, which are courts of the church whose functions are executive. These courts, composed of all the ministers of an assigned district, with a representative ruling elder from each sanctioned congregation, supervise all the churches within their bounds. The Ayrshire Presbyteries are subject to the Synod of Glasgow and Ayr, which, in turn, is subordinate to the General Assembly of the church.

The river Irvine divides the county into two parliamentary divisions, North and South Ayrshire, each of which returns one member to Parliament. Ayr and Kilmarnock are each the head of a group of parliamentary burghs returning a member. Irvine, one of the Ayr group, is the only additional burgh in the county with this privilege.

25. The Roll of Honour.

Ayrshire has produced men of mark in almost every sphere. Among her "worthies" are included men of letters, princes of industry, churchmen, lawyers, soldiers and statesmen, and men who took a large share in the struggles for civil and religious freedom.

Walter Kennedy, early Scottish poet, was born in one of the Carrick castles about 1460. He is best known as joint author of the *Flyting between Dunbar and Kennedy*. Dunbar commemorates him in his *Lament for the Makaris*. Alexander Montgomery, born at Hessilhead Castle near Beith, his paternal home, was another early Scottish poet,

James Boswell

whose fame rests on *The Cherrie and the Slae*. The county allows a high place among her men of letters to James Boswell of Auchinleck, the biographer of Samuel Johnson. Proudest, of course, of all Ayrshire's memories is that of Robert Burns, who stands unrivalled "for the mere essence of poetry and spirit of song." James Montgomery, writer of a large amount of deeply religious poetry, is connected with the county only by birth. He was born at Irvine, which he left at the age of four

Robert Burns

years. John Galt, Scottish novelist, was a native of the same town, and the district furnished much of the material on which his fame now rests. In depicting life in small Scottish towns, as in his masterpiece *The Annals of the Parish*, Galt is without a rival for rich humour, genuine pathos, and a rare mastery of the Scottish dialect. The Mures of Caldwell are a family known to fame in the learned and literary world. William Mure, Member of Parliament for Renfrewshire 1846–55, Lord Rector of Glasgow University 1847–48, was the author of *A Critical Account of the Language and Literature of Greece*, as well as of a *Journal of a Tour in Greece.*

In the long list of eminent statesmen and soldiers the name of Dalrymple cannot fail to find a place. The family, latterly taking its title from Stair, the Ayrshire village, has held a high place in the councils and activities of the country since the time when one of that house signalized himself among the Lollards of Kyle. In their time various members of this stock have taken a prominent share in several departments of public life,—as President of the Court of Session and Privy Councillor, as devotee of the Covenant and fierce disputant with Claverhouse, as Secretary of State sharing the infamy of the Glencoe massacre, or as aide-de-camp to Marlborough participating in the victories of the campaign in Flanders. Intimately connected with the county, and foremost in the public movements of his time, was Sir John Campbell, head of the house of Loudoun. He stoutly resisted Charles I in his attempt to force Episcopacy on Scotland. He also held the office of Lord High Chancellor of Scotland for nineteen years until he was deposed at the Restoration, was First Commissioner of the Treasury, and in 1648 was President of the Scottish Parliament. The turf and field-sports had a well-known patron in Archibald William, thirteenth Earl of Eglinton, who twice held the post of Lord Lieutenant of Ireland with great distinction and much popularity. He was also a Privy Councillor and Lord Rector of Glasgow University. But he is perhaps specially remembered among the people

General Neill

for his splendid reproduction of a tournament at Eglinton Castle in 1839, representing the chivalry of the past. Amongst the mailed knights who entered the lists to compete for the hand of the Queen of Beauty was Louis Napoleon, afterwards Napoleon III. The ancient seat of Kilkerran, in the parish of Dailly, has been rendered illustrious by Sir James Fergusson, the sixth baronet, born in 1832 and killed in a West Indian earthquake in 1907. He won renown by turns as soldier, colonial

The Earl of Dundonald

governor, and minister of state. The army and the navy have gained distinction from two of Ayrshire's sons. General Neill, of Swindrigemuir near Dalry, avenged the massacre of Cawnpore and fell in the advance on Lucknow. Thomas Cochrane, tenth Earl of Dundonald, made himself famous in the navy under the title of Lord Cochrane.

The church is represented by such names as Gavin Dunbar, born at New Cumnock, Dean of Moray, Archdeacon of St Andrews, Clerk-Register and Privy Councillor to James IV, and Bishop of Aberdeen, where he was a munificent benefactor; John Welsh, son-in-law to John Knox, and minister in the old church of Ayr; Robert Baillie, who, after receiving episcopal ordination, became parish minister of Kilwinning, joined the covenant against episcopacy, served as chaplain in the covenanting army at Duns Law, was appointed Professor of Divinity at Glasgow, and after the Restoration became principal of its University; Dickson of Irvine, who took a prominent part in the same stirring events; John Witherspoon, a lineal descendant of John Knox, who was minister at Beith during the rebellion of 1745, when he raised a band of volunteers for the service of George II, and gallantly fought at the battle of Falkirk, where he was taken prisoner, was some time minister at Paisley, went to America and became president of the College at Princeton; Andrew Kennedy Hutchison Boyd ("A. K. H. B."), who was born in Auchinleck Manse, and, after studying for the English Bar, ultimately settled as parish minister at St Andrews in 1865, became known by his essays in *Frazer's Magazine*, which were afterwards reprinted as "The Recreations of a Country Parson."

The Scottish Bar has been adorned by two Ayrshire names in addition to those already mentioned, the Right Honourable David Boyle, of Shewalton House, who held the position of Lord Justice General of Scotland, and John (Lord) Cowan, who became Solicitor General and Lord of Session.

Mention must be made of John Loudon McAdam, the inventor of the "macadamising" system of road-

William Murdoch, Inventor of Gas Lighting

making, who was a native of Ayr; William Murdoch, inventor of coal-gas, who was born between Lugar and Auchinleck; and Lord Kelvin, the world-renowned scientist, who for 30 years had his home at Netherhall, Largs. Alexander Macmillan, one of the founders of the great publishing house, was born and educated in Irvine. David Dale, who made a fortune by cotton-spinning and spent his last years in works of benevolence, belonged to Stewarton; and two Lord Mayors of London, Sir Andrew Lusk and Sir James Shaw, were natives of the county.

To Ayrshire belonged many of the martyrs of the wild times, when religious freedom had to be fought for and won through great sacrifice and suffering. Among these were John Brown, the Christian carrier of Priesthill, shot in the presence of his wife and children ; John Paton, Fenwick, who was captain in the army of Gustavus Adolphus, fought at Rullion Green and Bothwell Bridge, and was executed in the Grassmarket of Edinburgh ; Alexander Peden, "the prophet," a native of Sorn, who, after being ejected from his ministry at New Luce in Galloway, wandered through Ayrshire for many years, preaching at conventicles and hiding in caves. The *Scots Worthies* of John Howie, farmer at Lochgoin, near Fenwick, piously chronicles the sufferings of these martyrs of the Covenant.

26. THE CHIEF TOWNS AND VILLAGES OF AYRSHIRE.

(The figures in brackets after each name give the population in 1901, and those at the end of each section are references to the pages in the text.)

Alloway (969) on the right bank of the "bonnie Doon," two miles south of the town of Ayr. In an "auld clay biggin" here

Burns's Cottage, Ayr

Burns was born on 25 January 1759. Adjoining the cottage is a museum containing numerous valuable manuscripts and relics.

Burns's Cottage

The "haunted Kirk," the scene of Tam o' Shanter's frolic, still stands a roofless ruin. The "auld brig" spans the Doon, and hard by is the Burns monument. (pp. 18, 102, 115.)

Annbank (1303) is a large mining village, three miles south-west of Tarbolton, occupying a lovely situation, surrounded with green, cosy, woodland scenery on the winding Ayr. (p. 20.)

Auld Brig o' Doon

Ardrossan (6077) is a sea-port and watering-place. It possesses an important shipping connection. Its harbour is one of the safest and most accessible on the west coast of Scotland. The town is well situated both for trade and pleasure. On a hill above the town, called the Castle Craigs, stands a fragment of Ardrossan Castle said to have been strategically surprised by Wallace. (pp. 8, 29, 33, 47, 59, 87, 88, 94, 103, 123, 133, 135.)

Auchinleck (2168) is a town 15 miles east of Ayr by rail. Near it is Auchinleck House (locally called *Place Affleck*), the seat of the Boswells. Beside the mansion Sir Alexander Boswell, son of Johnson's biographer, established in 1815 the *Auchinleck Press* for printing MSS. of rare works in early English and Scottish literature. (pp. 30, 82, 134, 140, 145, 146.)

Ayr (29,101) is the county town standing on the bright sandy shore of Ayr Bay. The part south of the river constitutes the royal burgh whose charter was granted by William the Lion in the year 1202. In Ayr was held the first parliament under Bruce in 1315, which settled the succession to the Scottish crown. The

"Twa Brigs," Ayr

river is spanned by five bridges, the best known being the "twa brigs" of Burns. The "auld brig," so narrow that "twa wheel-barrows tremble when they meet," owes its origin probably about the middle of the thirteenth century to two philanthropic ladies who, being grieved at the spectacle of so many people being drowned at the ford, resolved thus to spend their wealth. It is a splendid example of ancient bridge architecture. Ayr is an important market town and the centre of an agricultural community long noted for its productions and prosperity. There are gathered here each April the finest examples of the world-famed Ayrshire cattle. Its numerous natural advantages of situation make it an attractive and well-favoured town. A splendid new water-supply, drawn from Loch Finlas 20 miles distant, was introduced in 1887; and the corporation has provided electric lighting and electric tramways for the town and vicinity. A fine esplanade

The Auld Brig o' Ayr

stretches south from the harbour, and a new race course was opened in 1907. General Neill of Swindrigemuir, one of the heroes of the Indian mutiny, and Archibald, Earl of Eglinton, "The Good Earl," are commemorated by monuments. There is also a handsome and artistic memorial of Burns. Ayr, Irvine, Campbeltown, Inverary, and Oban form what is called the Ayr group of burghs, and unite in sending one member to parliament. (pp. 2, 20, 49, 59, 65, 73, 75, 87, 90, 94, 96, 101, 111, 114, 116, 123, 133, 134, 135, 136, 137, 138, 139, 145, 146.)

Ballantrae (511) is a village situated at the mouth of the Stinchar. Its ancient name was Kirkcudbright-Innertig (church of St Cuthbert standing at the influx of the Tig). It is principally supported by its herring and salmon fisheries. Formerly it was notorious as a smuggling centre; within recent years it has become a popular summer resort with many attractions. (pp. 15, 33, 52, 86, 134.)

Barassie is a favourite summer retreat one and a half miles north of Troon. Large engineering works connected with the Glasgow and South Western Railway workshops have been recently erected south of the hamlet. (pp. 49, 135.)

Barr (581) is a pastoral village nestling between the Stinchar and its tributary the Gregg and fully five miles from a railway station. Its air is invigorating, and the mineral waters of Shalloch Well in past years used to invite to the district people of rank and fashion. About one and a half miles below the village crowning a bluff are the ruins of Kirk Dominae. (p. 66.)

Barrhill (302) is a quiet moorland village on the Dusk, a large tributary of the Stinchar. Its chief interest lies in its proximity to the *Raider* country. The drive to Glen Trool in the mountainous region of the Merricks is through wild and interesting scenery.

Barrmill (635), a village south-east of Beith, has a factory for linen thread. (p. 77.)

Beith (4963), on an eminence in the northern extremity of the county, commands an extensive view of the country around and of Kilbirnie Loch. About 1760 the making of white thread was begun, but towards the end of the century it declined. Tanning, currying and preparation of leather are now carried on together with rope spinning and the production of fishing and other nets. But the staple industry is represented by several extensive up-to-date factories for the production of high-class furniture of every description. A mile to the north of the town, at Roebank, silk-printing is carried on. The neighbourhood produces coal and iron, and limestone, though abundant, is burnt now only to a limited extent. Spier's School is a seminary for secondary education built in 1887 after the model of the old college of Glasgow. (pp. 13, 29, 30, 32, 38, 72, 74, 79, 82, 83, 99, 106, 109, 111, 121, 133, 135, 140, 145.)

Catrine (2340) is a town on the Ayr. It is compactly built on a modern plan and finely sheltered on almost all sides—by the woods of Catrine Bank on the east, by those of Ballochmyle on the north, and by plantation belts on the south. It was founded in 1785 by the celebrated philanthropist David Dale in company with the first Claud Alexander of Ballochmyle, who established extensive cotton works here. The motive power of the factory is mainly supplied by the river which drives two enormous water wheels. (pp. 20, 77.)

Colmonell (306) is a village on the right bank of the Stinchar three miles west of the railway at Pinwherry. (pp. 102, 119.)

Cumnock, New (2005) is a pleasant upland village, 800 feet above sea-level, widely scattered about the confluence of the Nith river, Afton Water, and Moorfoot Burn. (pp. 18, 30, 66, 81, 83, 84, 102, 107, 134, 135, 145.)

Cumnock, Old (3088) is a police burgh situated 18 miles east of Ayr in a deep sheltered hollow at the confluence of the Glaisnock and Lugar Waters. It is in the heart of a busy mining and agricultural district, and mining is now the chief industry. But woollen manufactures, wool spinning and dyeing, coach-building, and iron-founding are also carried on. There are also engineering works for the production of threshing-mills. The manufacture of wooden snuff-boxes was many years ago transferred to Mauchline. Two miles west of the town is Dumfries House, a seat of the Marquis of Bute, and in its grounds is Terringzean Castle now in ruins. The environs are marked by singular beauty and variety, which are much intensified by two lofty railway viaducts—one over the picturesque valley of the Glaisnock, and the other over a deep chasm through which flows the Lugar. (pp. 66, 77, 82, 102, 123, 134.)

Dailly (502) is a village in a low but charming situation, six miles from Maybole. Coal and iron abound in the neighbourhood and tile-works are near. In the neighbourhood are Dalquharran Castle and Kilkerran House. The proprietors of both these estates a century ago did much to beautify the district by afforestation. (pp. 16, 31, 81, 116, 144.)

Dalmellington (1448) stands on the Doon 15 miles from Ayr. The village is ancient, and has many historical and antiquarian associations. The district is the richest in the county for mineral products. About three miles north-west are the extensive ironworks and collieries of the Dalmellington Iron Company, employing some 2000 men. (pp. 11, 18, 28, 82, 101, 102, 134.)

Dalry (5316) is a town in the north of the county pleasantly situated on the Garnock. Formerly the chief employment was silk and cotton hand-loom weaving, but this has now given way to the raising of coal, ironstone, and limestone found in the neighbourhood. The other industries comprise worsted spinning, the manufacture of blankets, tweeds and hosiery, cabinet-making and brick-making. (pp. 22, 29, 30, 63, 77, 82, 83, 104, 106, 123, 129, 133, 134, 135, 144.)

Dalrymple (305) is a village beautifully situated on the Doon four and a half miles from Ayr. A considerable number of its inhabitants are engaged in the manufacture of blankets at Skeldon Mills. (pp. 18, 63, 135.)

Darvel (3070) is a police burgh nine miles east of Kilmarnock on the Irvine. The manufacture of tapestry curtains, carpets, lace, muslin and other light fabrics affords employment to a large number of people. (pp. 77, 134.)

Dreghorn (1155) is pleasantly situated on a gentle eminence descending to the west two miles from Irvine. (pp. 33, 82, 83, 134.)

Dunaskin or **Waterside** (1407) is a village near Dalmellington. It was founded in 1847 along with the extensive Dalmellington Ironworks adjacent. (pp. 18, 134.)

Dundonald (339) is a village five and a half miles from Kilmarnock amidst cultivated fields and wooded heights. The village and surrounding property came into the hands of Sir William Cochrane, who obtained a charter from Charles I in 1638 creating the Kirkton of Dundonald into a free burgh of barony, a privilege that it has never exercised. At the Restoration the title of the superior was raised to that of Lord Cochrane, Earl of Dundonald. (p. 104.)

Dunlop (473) is a village on the Glazert burn, eight miles from Kilmarnock. It has long been noted for the excellent quality of its cheese, which bears a high character throughout Scotland. The inhabitants of the village and surrounding district carry on a considerable trade in meal and in ham-curing (Ayrshire bacon). Near the village and resting on a small knoll is a stone of large dimensions called the "O great Stone," supposed to have been employed in some ancient rites of worship. (pp. 69, 116, 134.)

Eglinton (1075) is a village one and a half miles south of Kilwinning with ironworks. (p. 80.)

Failford is a small hamlet romantically situated at a point where the Fail Water joins the Ayr. It is famous for its Water-of-Ayr stones or hones.

Fairlie (674) is a watering-place on the Clyde two and a half miles from Largs. It is celebrated for yacht-building, and from its yards have come many famous boats including several American Cup challengers. (pp. 29, 45, 55, 83, 135.)

Fairlie, near Largs

Fenwick (329) is a village four and a quarter miles from Kilmarnock. Lochgoin farmhouse, far away on the wild moors near the north-east boundary of the parish, five or six miles from Fenwick by road, was the birthplace and residence of John Howie author of the *Scots Worthies*. Many memorials of the covenanters are preserved in the farmhouse including the flag of the Fenwick covenanters employed at the encounter of Bothwell Bridge together with the sword and bible of Captain Paton. (pp. 133, 147.)

On the River at Girvan

Galston (4876) is a police burgh, occupying a sheltered position on the left of the Irvine Water. It was formerly a burgh of barony, the Duke of Portland delegating to the people his prerogative of governing the place. Its main industries are weaving, lace and muslin making, and coal-mining. The churchyard is hallowed with the tombs of several martyrs. (pp. 77, 82, 84, 98, 100, 131, 134.)

Girvan (4024) is a police burgh, a sea-port, a fishing station, and a watering place at the mouth of the river of the same name.

It had an existence as early as the eleventh century. In 1668 the feudal superior with visionary schemes procured a patent for building a new burgh. The actual prosperity of the place dates from the introduction of hand-loom weaving, of which it was a centre. Of late years Girvan has become a popular resort for summer visitors. It is the centre of a productive agricultural district, where the cultivation of early potatoes is prosecuted on scientific principles. (pp. 9, 16, 33, 37, 51, 56, 59, 69, 84, 86, 87, 102, 111, 129, 134, 135.)

Irvine, from River

Glenbuck (1037) is a thriving village three and a half miles east of Muirkirk and half a mile from the most easterly point of the county. Mining and ironworks furnish the chief industries. (p. 20.)

Glengarnock (2087) is a progressive place between the south end of Kilbirnie Loch and the river Garnock with large iron, steel and chemical works, employing over 2000 persons (p. 80.)

Hurlford (4205) is a town on the Irvine two miles from Kilmarnock with ironworks, collieries, and fire-clay works. (pp. 33, 80, 83, 134.)

Irvine (9618) is a royal burgh, and a flourishing town and sea-port on the north bank of the river of the same name. It ranks as one of the oldest of the Scottish royal burghs, having received its charter early in the thirteenth century. During the seventeenth and eighteenth centuries it held a prominent position as a shipping port, being largely used by Glasgow merchants for sea-borne trade before the construction of the harbour at Port-Glasgow. The harbour has been much improved at very considerable expense and the trade has greatly increased especially in the export of coal and the import of limestone. Its prosperity is also due in no small degree to extensive chemical and chrome works on the south side of the harbour. Some of the notables associated with the town in the past were Lord Justice General Boyle, Eckford the designer of the American navy, John Galt, author of the *Annals of the Parish*, James Montgomery, a minor poet, Burns, who worked for some time as a flax-dresser, and Elspet Simpson, "Luckie Buchan," founder of the Buchanites, an extraordinary sect of fanatics. Irvine joins with the other members of the Ayr burghs in sending one member to parliament. (pp. 13, 21, 22, 31, 49, 59, 77, 87, 90, 94, 99, 111, 114, 120, 123, 133, 134, 135, 136, 139, 141, 145, 147.)

Kilbirnie (4571) stands on the Garnock nine miles from its mouth. The principal industries are flax-spinning, weaving of fishing-nets, and the manufacture of linen thread. Objects of interest are the parish church, the old Castle of Glengarnock, and the ruins of Kilbirnie Castle. (pp. 79, 82, 83, 93, 106, 109, 116, 122.)

Kilbride, West (2315) is a town four and a quarter miles from Ardrossan. It is a favourite summer resort. A monument 50 feet high commemorates Dr Robert Simson, for 50 years

Professor of Mathematics in the University of Glasgow, and a native of the parish. Off the coast one of the largest vessels of the Spanish Armada remains sunk in 10 fathoms of water. About two centuries ago an attempt was made to examine the internal condition of the ship with the result that a piece of ordnance was raised from the hulk. A reference to this operation is found in Defoe's *Tour through Great Britain*. The cannon still lies on the

Laigh Kirk and Steeple (Kilmarnock)

green beside Portincross Castle. (pp. 4, 29, 30, 31, 32, 47, 69, 83, 103, 106, 116, 134, 135.)

Kilmarnock (35,091) is the largest town in the county standing at the confluence of the Kilmarnock Water with the river Irvine. About 1200 a church was dedicated to an Irish saint of the seventh century. The town was made a burgh of barony by James VI in 1591, a royal burgh by Charles II in 1672, and a parliamentary burgh in 1832. The Laigh Kirk still retains its old spire erected in 1410. The town seems to have been early distinguished for the manufacture of hose, Scotch bonnets, and military caps. In the Kay Park is a Burns monument; and in the centre of the town is a statue to the memory of Sir James Shaw. From Kilmarnock House in Marnock Street, the last residence of the noble family of Kilmarnock, the fourth earl proceeded in 1745 to join the standard of the Pretender. He was taken

Council House, Kilmaurs

prisoner at Culloden and executed on Towerhill. Kilmarnock played an important part in the political movement known as Chartism. Along with Renfrew, Rutherglen, Dumbarton, and Port-Glasgow it unites in sending one member to parliament. (pp. 13, 33, 66, 70, 75, 77, 82, 83, 95, 111, 116, 121, 123, 128, 133, 134, 135, 136, 138, 139.)

Kilmaurs (1803) is a town and burgh two and a quarter miles from Kilmarnock. It was erected into a burgh of barony in 1527 by charter from James V. It was at one time famous for its cutlers, whose blades were famous, as would appear from the current proverbial expression "gleg as a Kilmaurs whittle." The principal industry now is the manufacture of boots, shoes, and hosiery. (pp. 82, 105, 116, 134, 136.)

Kilwinning (4440) is a police burgh pleasantly situated on the Garnock. The trade consists partly in the weaving and manufacture of gauzes, muslins, and shawls, and a large woollen factory has been recently erected. In the neighbourhood are large iron, coal, and fire-clay works, as well as several engineering works and iron foundries. Kilwinning is famous as the traditional birth-place of freemasonry in Scotland, introduced, it is supposed, by the foreign architect engaged in the erection of the abbey. It is also remarkable for the perpetuation of ancient Scottish archery established in 1488, and, after a discontinuance and revival, continued down till 1870. The shooting at the popinjay (papingo), placed in the steeple 105 feet high, is described in Scott's *Old Mortality*. (pp. 13, 33, 80, 82, 83, 111, 116, 133, 135, 145.)

Kirkoswald is a village four and a half miles from Maybole. In the neighbourhood are the farms of Shanter and Jameston, the former the residence of Douglas Graham and the latter of John Davidson, shoemaker—the prototypes of the famous "Tam o' Shanter and Souter Johnny" of Burns. The graves of both these worthies are pointed out in Kirkoswald churchyard, beside those of a number of the poet's maternal ancestors. (pp. 106, 116.)

Largs (3246) is a police burgh and watering place opposite the north end of Great Cumbrae. Tradition has it that a mission church was founded here by St Columba, the celebrated Abbot of Iona. From this circumstance he is recognized as the patron saint of the place, and St Columba's (Colm's) day is an annual fair and festival held in his honour. Largs owes its historical fame to a battle fought here between the Scots and Norwegians in 1263. (pp. 29, 45, 93, 111, 133, 135, 146.)

Loans is a village of some antiquity one and a half miles east of Troon on the Dundonald road. In former times it was a hiding place for smugglers, and the memory of contraband exploits is perpetuated in some of the local names. An old house on the Irvine road is called "Rum How" while a field near is called "Brandy Hill," and several "Brandy Holes," excavations beneath the dwellings, have been recently disclosed.

Loudounkirk

Loudoun (Loudounkirk) is a hamlet two and a half miles west of Newmilns on the Irvine. The chancel of the old parish kirk, the only part now remaining, is used as the burying place of the family of Hastings (Loudoun Castle).

Lugar (1286) is a village one and a half miles from Cumnock. Large ironworks adjoin the village. (p. 80.)

Maidens is a quiet fishing retreat one mile north of Turnberry. (pp. 50, 56.)

Mauchline (1767) stands 17½ miles from Ayr. The history of the town begins in 1165 when the lands of Mauchline were gifted by Walter the High Steward to the monks of Melrose, who are generally believed to have planted here a colony of their order and to have built a church. Mauchline is associated with the life and works of Burns. The old kirk was the scene of his *Holy Fair*, and in the graveyard may be seen the tombstones of Mary Morrison, "Daddy Auld," "Nance Tannoch," "Holy Willy," and

Ballochmyle Viaduct

"Racer Jess." Opposite the churchyard gate is the cottage of "his Jolly Beggars." Mauchline has long been noted for its manufacture of wooden snuff-boxes. and numerous other knickknacks beautifully artistic in design and finish. South-east of the town is the great Ballochmyle quarry of red sandstone and near it is a splendid viaduct over the Ayr. The chief arch is 180 feet wide, while the height of the parapet above the gushing water is 196 feet.

Three smaller arches fill up the ends till the sylvan-fringed banks are reached at the railway level. (pp. 30, 63, 73, 83, 114, 134, 135.)

Maybole (5892) is a police burgh eight and a half miles from Ayr. In the sixteenth and seventeenth centuries Maybole was a place of considerable importance, and, as the capital of Carrick, it once boasted of 28 baronial mansions. Formerly, the chief

Howford Bridge, near Mauchline

industry was hand-loom weaving; now its inhabitants are mainly employed in boot and shoe-making and the manufacture of agricultural implements. (pp. 32, 99, 105, 115, 116, 117, 134, 135, 136.)

Muirkirk (3892) is a town standing on the river Ayr four miles from its source. It is the most easterly town in Ayrshire and is situated 720 feet above sea-level. The *moor kirk* was erected in 1650. Before that time the place was included in the parish of Mauchline and called the Garron. Two miles east of the town on the farm of Priesthill a monument has been erected to the memory of John Brown the Covenanter. The famous ironworks were commenced in 1787, from which time the town has gradually expanded. Three blast and several puddling furnaces for the manufacture of pig and malleable iron besides extensive works for the manufacture of chemicals are in constant operation. Coal and limestone are abundant in the vicinity. (pp. 7, 20, 32, 38, 79, 82, 134.)

Newmilns (4467) is a thriving police burgh on the Irvine seven and a half miles from Kilmarnock. Along with Galston and Darvel in the same valley it is a seat of the lace curtain industry which is carried on extensively. (pp. 77, 98, 134, 136.)

Newton-on-Ayr is a suburb of Ayr on the right bank of the river. Until recently it formed a separate burgh. Burghal privileges were conferred on it by King Robert Bruce in 1315.

Ochiltree (549) is a quiet agricultural village 11½ miles east of Ayr. It consists mainly of one street half a mile in length. The village is the scene of *The House with the Green Shutters* by George Douglas, a native of the place. (pp. 11, 70, 105, 120.)

Patna (482) is a quaint old-fashioned mining village on the Doon five miles north-west of Dalmellington. It has some cotton works, and coal and iron are plentifully obtained in the neighbourhood. (pp. 18, 77, 134.)

Prestwick (2766) stands on the coast near Ayr. Its ruined church is said to date from 1163. It was erected into a burgh of barony in the twelfth century, and from its charter, renewed by James VI in 1600, it was granted the privilege of holding a weekly market and an annual fair. The market cross of considerable antiquity still stands. To-day Prestwick is one of the most modern of sea-bathing towns, and is the head-quarters of golf on the Ayrshire coast. (pp. 49, 80, 83, 115, 116, 123, 133, 136.)

Rankinstone (667) is a moorland mining village in the parish of Coylton four miles north-east of Patna.

Riccarton on the south bank of the Irvine Water is a suburb of Kilmarnock, with which it is connected by a bridge. A considerable quantity of coal is raised in the neighbourhood. Caprington Castle in the vicinity is an ancient and historic building. (pp. 82, 93.)

St Quivox (St Kevoch), a village on the Ayr, has coal-mines and excellent sandstone quarries. Near it are the fine wooded glades of Auchincruive.

Saltcoats (8120) is a police burgh and watering place near Ardrossan. It was a seat of salt manufacture from 1686 to 1827. It is a clean old-fashioned town mainly built on a slight promontory, two nebbocks, pointing seawards. Some of its streets are quaint and narrow. It possesses a picturesque old pier used only by fishermen, the only instance of its kind on the west coast. The harbour used to be of some importance and had considerable trade with Ireland. This began to decline in 1837 on account of the growth of its neighbouring rival, Ardrossan. The town is a popular sea-bathing resort. In 1495 the Earl of Glencairn, the superior, granted for 999 years to nine fishermen of Saltcoats leases of land known as the "nine yards" on condition that every spring they in their two boats carried the Earl's furniture from the creek of Saltcoats to Finlayston, and brought it back in the late autumn when the family returned to their residence at Kerilaw. Both the "nine yards" and the "cots" of the saltmakers are merely antiquarian relics. (pp. 47, 80, 111, 133.)

Seamill is on the outskirts of West Kilbride, a short distance to the west. It has a fine stretch of sand suitable for bathing, and a golf-course stretches north along the shore. A fort of the late Celtic period, when explored, yielded wheel-shaped pendants and other relics.

Skelmorlie (1096) is a fashionable watering-place and summer resort at the northern extremity of the county. It came into existence about 1850. The cliff and fine sea-beach make the place a desirable retreat. (pp. 29, 32, 42, 133.)

Sorn, a village three and three-quarter miles east of Mauchline, lies on the right bank of the river Ayr in a beautiful holm almost enclosed with woods. (pp. 20, 82, 134, 147.)

Stevenston (6554) is a town one mile from Saltcoats. It is an ancient place, St Monoch's Fair being mentioned in a charter of 1189. The staple industries of the town used to be cotton and silk weaving. It now depends on ironworks, chemical works, foundries, collieries, and brickworks; and Nobel's Explosives factory is in the neighbourhood. (pp. 31, 33, 79, 80, 82, 83, 133.)

Stewarton (2858) is a town five and a half miles from Kilmarnock. Its specialty is the manufacture of Scotch bonnets, but it also carries on carpet-weaving, spindle-making for cotton and woollen mills, carding and spinning, Ayrshire needlework, and the manufacture of steel clock-work. (pp. 13, 82, 106, 134, 147.)

Straiton is a village prettily situated among hills, six and a half miles from Maybole. Pastoral hills surround it on the east, south, and west, and the broad cultivated vale of the Girvan lies on the west. It has some manufactures of cottons and tartans. (pp. 103, 130.)

Symington is a village five miles from Kilmarnock. Freestone and whinstone underlie the surface of the district round. In the vicinity are the finely wooded grounds of Coodham, Dankeith, and Rosemount. (pp. 31, 83, 134.)

Tarbolton (926) is a village seven miles from Ayr. The name, "the beltane hill," suggests that it was at one time the seat of fire-worship. The hill on which the fire is said to have been lit is that close to the Mauchline road as it leaves the village. The practice of lighting a fire there still survives although its meaning is forgotten; and on the eve of St John in the month of June boys go from house to house demanding fuel for the bonfire. The same practice is perpetuated in other places near, as at Prestwick. Tarbolton has many associations with Burns. About a quarter of a mile from the village on the road to Lochlee and Mossgiel is Tarbolton Mill, "Willie's Mill," the scene of ***Death and Dr Hornbook***. One mile south-east is Coilsfield —the "Castle o' Montgomery"—the scene of one of the poet's happiest lyrics. (pp. 82, 98, 105, 116.)

Troon (4764) is a police burgh, sea-port, and watering-place, six miles north of Ayr. The rocky promontory, a quarter of a mile broad, on which a considerable part of the town is built, extends fully a mile into the sea. Troon has several golf-courses, and splendid sands. (pp. 7, 8, 47, 49, 56, 59, 80, 87, 90, 99, 111, 135.)

Waterside, see **Dunaskin.**

West Kilbride, see **Kilbride.**

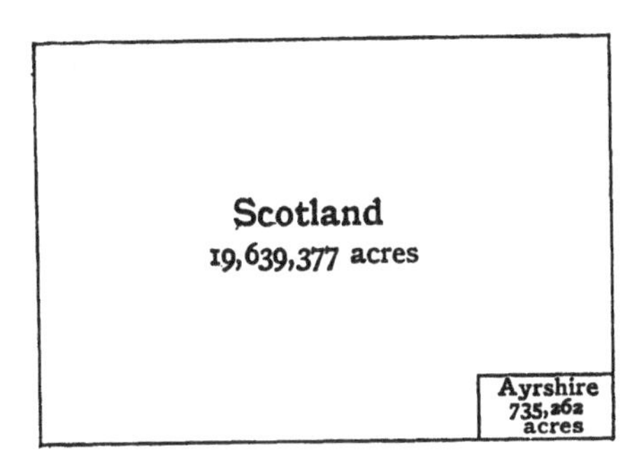

Fig. 1. Area of Ayrshire compared with that of Scotland

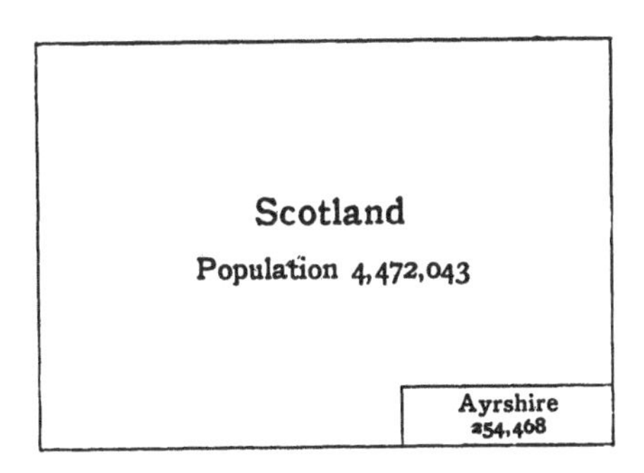

Fig. 2. The population of Ayrshire compared with that of Scotland

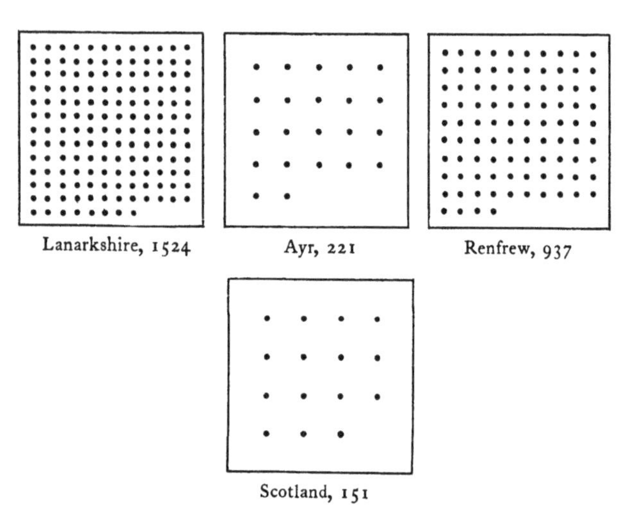

Fig. 3. Comparative density of Population to the Square Mile (1901)

(*Note—Each dot represents* 10 *persons*)

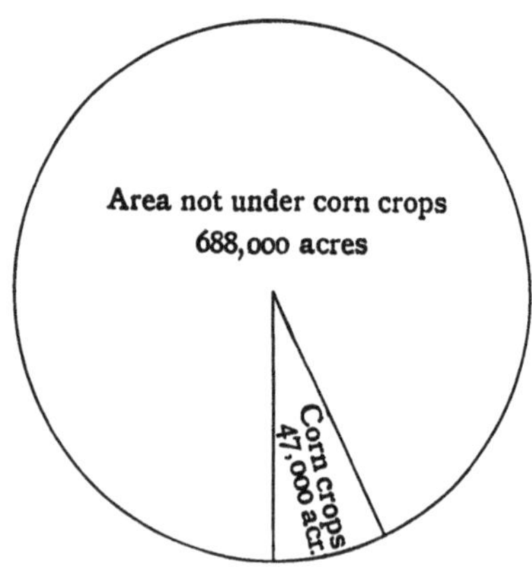

Fig. 4. Proportionate area of Ayrshire under Corn Crops

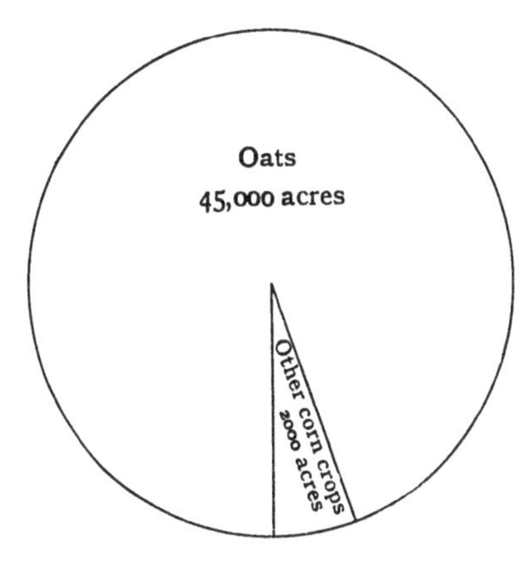

Fig. 5. Proportionate area of cultivation of oats and other Corn Crops in Ayrshire

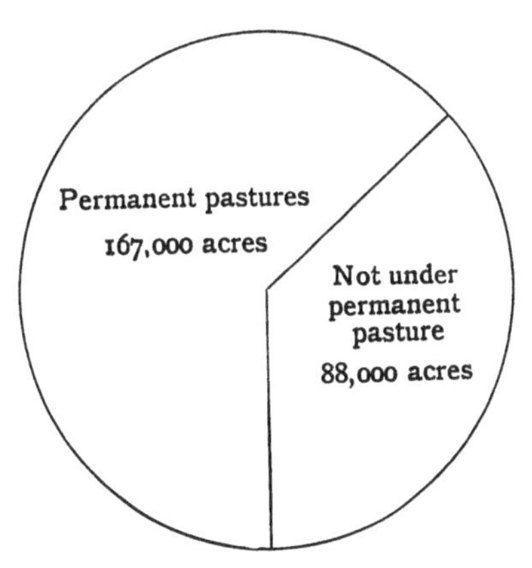

Fig. 6. Proportionate areas of grass-lands in Ayrshire under Permanent Pasture and not under Permanent Pasture

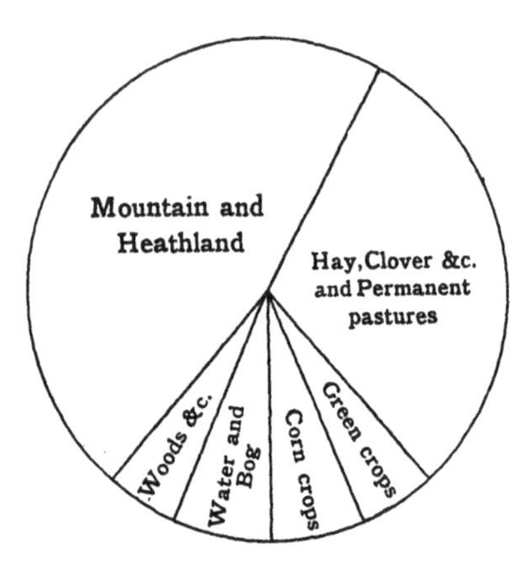

Fig. 7. Proportionate areas of arable and non-arable lands in Ayrshire

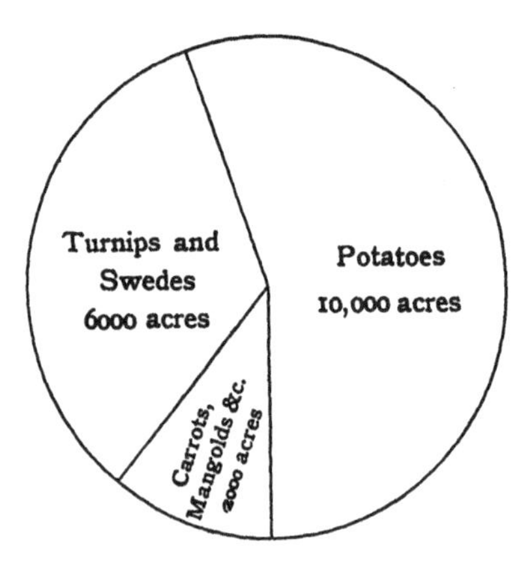

Fig. 8. Proportionate areas under Green Crops in Ayrshire

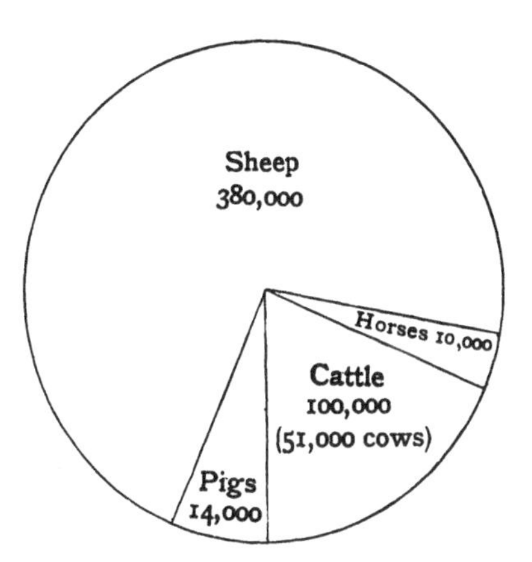

Fig. 9. Proportionate numbers of Live Stock in Ayrshire

www.ingramcontent.com/pod-product-compliance
Ingram Content Group UK Ltd.
Pitfield, Milton Keynes, MK11 3LW, UK
UKHW042144280225
455719UK00001B/79

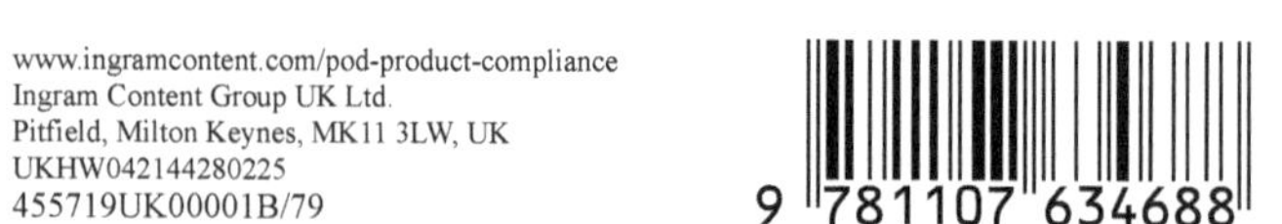